AF594942

Tactile Sensing Technology and Systems

Tactile Sensing Technology and Systems

Special Issue Editor

Maurizio Valle

MDPI • Basel • Beijing • Wuhan • Barcelona • Belgrade • Manchester • Tokyo • Cluj • Tianjin

Special Issue Editor
Maurizio Valle
University of Genova
Italy

Editorial Office
MDPI
St. Alban-Anlage 66
4052 Basel, Switzerland

This is a reprint of articles from the Special Issue published online in the open access journal *Micromachines* (ISSN 2072-666X) (available at: http://www.mdpi.com).

For citation purposes, cite each article independently as indicated on the article page online and as indicated below:

LastName, A.A.; LastName, B.B.; LastName, C.C. Article Title. *Journal Name* **Year**, *Article Number*, Page Range.

ISBN 978-3-03936-501-2 (Pbk)
ISBN 978-3-03936-502-9 (PDF)

Contents

About the Special Issue Editor

Maurizio Valle received the M.S. degree in Electronic Engineering in 1985 and the Ph.D. degree in Electronics and Computer Science in 1990 from the University of Genova, Italy. In 1992 he joined the University of Genova, first as an assistant and in 2007 as an associate professor. From December 2019, MV is full professor at the Department of Electrical, Electronic and Telecommunications Engineering and Naval Architecture, University of Genova and he is Head of Connected Objects, Smart Materials, Integrated Circuits – COSMIC laboratory. MV has been and is in charge of many research contracts and he is co-funder of two spin-offs. MV is co-author of more than 200 articles on international scientific journals and conference proceedings. His research interests include electronic and microelectronic systems, material integrated sensing systems, tactile sensors and electronic-skin systems, wireless sensor networks. He is IEEE senior member and member of the IEEE CAS Society.

Editorial

Editorial of Special Issue "Tactile Sensing Technology and Systems"

Maurizio Valle

Department of Electrical, Electronic, Telecommunications Engineering and Naval Architecture, University of Genova, Via Opera Pia 11A, I16145 Genova, Italy; maurizio.valle@unige.it

Received: 12 May 2020; Accepted: 14 May 2020; Published: 16 May 2020

Human skin has remarkable features such as self-healing ability, flexibility, stretchability, high sensitivity and tactile sensing capability. It senses pressure, humidity, temperature and other multifaceted interactions with the surrounding environment. The imitation of human skin sensing properties via electronic systems is one of the frontrunner research topics in prosthetics, robotics, human-machine interfaces, artificial intelligence, virtual reality, haptics, biomedical instrumentation and healthcare, to name but a few.

Electronic skins or artificial skins are devices that aim to assimilate and/or mimic the versatility and richness of the human sense of touch via advanced materials and technologies. Generally, electronic skins encompass embedded electronic systems which integrate tactile sensing arrays, signal acquisition, data processing and decoding.

Tactile sensors sense diversity of properties via direct physical contact (i.e., physical touch), e.g., vibration, humidity, softness, texture, shape, surface recognition, temperature, shear and normal forces. Tactile sensors are dispersed sensors that translate mechanical and physical variables and pain stimuli into electrical signals. They are based on a wide range of technologies and materials, e.g., capacitive, piezoresistive, optical, inductive, magnetic and strain gauges. Artificial tactile sensing allows the detection, measurement and conversion of tactile information acquired from physical interaction with objects. In the last two decades, the development of tactile sensors has shown impressive advances in sensor materials, design and fabrication, transduction techniques, capability and integration; however, tactile sensors are still limited by a set of constraints related to flexibility, conformability, stretchability, complexity and by real-time implementation of information decoding and processing.

The Special Issue collects eight published papers, tackling the fabrication, integration and implementation of tactile sensing in some of the abovementioned applications such as haptics, robotics, human-computer interaction and artificial intelligence, modeling, decoding and processing of tactile information using machine learning techniques.

Particularly, Wang et al. presented a flexible tactile sensor array (3×3) with surface texture recognition method for human-robotic interactions. They developed and tested a novel method based on Fine Element Modeling and phase delay to investigate the usability of the proposed flexible array for slippage and grooved surface discrimination when slipping over an object [1]. Choi et al. developed a skin-based biomimetic tactile sensor with bilayer structure and different elastic moduli to emulate human epidermal fingerprints ridges and epidermis. They proved that the proposed sensor has a texture detection capability for surfaces under 100 μm with only 20 μm height difference [2]. Chen et al. investigated the influence of skin thickness and thermal contact resistance on a thermal model for tactile perception. They proposed and tested a novel methodology to reproduce thermal cues for surface roughness recognition [3]. Kameoka et al. developed and assessed a pressure distribution sensor that measures stickiness when touching an adhesive surface via magnetic force offset [4]. Cordoba et al. proposed a set of calibration methods, Quasi Single Point Calibration Method (QSPCM), other than two-point calibration method (TPCM) for high-speed measurements of resistive sensors. The FPGA implementation of the proposed circuit has been used to quantify resistances values in the range (267.56

W,7464.5 W) [5]. Byun et al. presented a new gesture recognition method implemented on Flexible Epidermal Tactile Sensor FETSA based on strain gauge to sense object deformation. They prototyped and implemented a wearable hand gesture recognition smart watch. The latter demonstrated its ability to detect eight motions of the wrist and showed higher performance than preexisting arm bands, in terms of robustness, stability and repeatability [6]. Maurel et al. developed a viso-tactile substitution system based on vibrotactile feedback called TactiNET for the active exploration of the layout and the typography of web pages in a non-visual environment, the idea being to access the morpho-dispositional semantics of the message conveyed on the web. They evaluated the ability of the TactiNET to achieve the categorization of web pages in three domains—namely tourism, e-commerce and news [7]. Finally, Alemeh et al. assessed a comparison of embedded machine learning techniques—specifically convolutional neural networks models—for tactile data decoding units on different hardware platforms. The proposed model shows a classification accuracy around 90.88% and outperforms the current state-of art in terms of inference time [8].

I would like to take the opportunity to give my genuine thanks to all the authors for submitting such valuable scientific contributions to this Special Issue. Furthermore, my sincere undisputed thanks to all the reviewers for dedicating time and effort to revising the various manuscripts.

References

1. Wang, Y.; Chen, J.; Mei, D. Flexible Tactile Sensor Array for Slippage and Grooved Surface Recognition in Sliding Movement. *Micromachines* **2019**, *10*, 579. [CrossRef] [PubMed]
2. Choi, E.; Sul, O.; Lee, J.; Seo, H.; Kim, S.; Yeom, S.; Ryu, G.; Yang, H.; Shin, Y.; Lee, S.-B. Biomimetic Tactile Sensors with Bilayer Fingerprint Ridges Demonstrating Texture Recognition. *Micromachines* **2019**, *10*, 642. [CrossRef] [PubMed]
3. Chen, C.; Ding, S. How the Skin Thickness and Thermal Contact Resistance Influence Thermal Tactile Perception. *Micromachines* **2019**, *10*, 87. [CrossRef] [PubMed]
4. Kameoka, T.; Takahashi, A.; Yem, V.; Kajimoto, H.; Matsumori, K.; Saito, N.; Arakawa, N. Assessment of Stickiness with Pressure Distribution Sensor Using Offset Magnetic Force. *Micromachines* **2019**, *10*, 652. [CrossRef] [PubMed]
5. Botín-Córdoba, J.A.; Oballe-Peinado, Ó.; Sánchez-Durán, J.A.; Hidalgo-López, J.A. Quasi Single Point Calibration Method for High-Speed Measurements of Resistive Sensors. *Micromachines* **2019**, *10*, 664. [CrossRef] [PubMed]
6. Byun, S.-W.; Lee, S.-P. Implementation of Hand Gesture Recognition Device Applicable to Smart Watch Based on Flexible Epidermal Tactile Sensor Array. *Micromachines* **2019**, *10*, 692. [CrossRef] [PubMed]
7. Maurel, F.; Dias, G.; Safi, W.; Routoure, J.-M.; Beust, P. Layout Transposition for Non-Visual Navigation of Web Pages by Tactile Feedback on Mobile Devices. *Micromachines* **2020**, *11*, 376. [CrossRef] [PubMed]
8. Alameh, M.; Abbass, Y.; Ibrahim, A.; Valle, M. Smart Tactile Sensing Systems Based on Embedded CNN Implementations. *Micromachines* **2020**, *11*, 103. [CrossRef] [PubMed]

Article

Layout Transposition for Non-Visual Navigation of Web Pages by Tactile Feedback on Mobile Devices

Fabrice Maurel [1,*], Gaël Dias [1], Waseem Safi [2], Jean-Marc Routoure [1] and Pierre Beust [1]

1 Groupe de Recherche en Informatique, Automatique, Image et Instrumentation (GREYC), National Graduate School of Engineering and Research Center (ENSICAEN), Université de Caen Normandie (UNICAEN), 14000 Caen, France; gael.dias@unicaen.fr (G.D.); jean-marc.routoure@unicaen.fr (J.-M.R.); pierre.beust@unicaen.fr (P.B.)

2 Higher Institute for Applied Sciences and Technology, Damascus, Syria; waseem.safi@hiast.edu.sy

* Correspondence: fabrice.maurel@unicaen.fr

Received: 8 January 2020; Accepted: 25 March 2020; Published: 3 April 2020

Abstract: In this paper, we present the results of an empirical study that aims to evaluate the performance of sighted and blind people to discriminate web page structures using vibrotactile feedback. The proposed visuo-tactile substitution system is based on a portable and economical solution that can be used in noisy and public environments. It converts the visual structures of web pages into tactile landscapes that can be explored on any mobile touchscreen device. The light contrasts overflown by the fingers are dynamically captured, sent to a micro-controller, translated into vibrating patterns that vary in intensity, frequency and temperature, and then reproduced by our actuators on the skin at the location defined by the user. The performance of the proposed system is measured in terms of perception of frequency and intensity thresholds and qualitative understanding of the shapes displayed.

Keywords: vibrotactile feedback; blind users; web accessibility

---

1. Introduction

Voice synthesis and Braille devices are the main technologies used by screen readers to afford access to information for visually impaired people (VIP). However, they remain ineffective under certain environmental conditions, in particular on mobile supports. Indeed, in 2017, the *World Health Organization* (WHO: https://www.who.int/) counted about 45 million blind people in the world. However, a study by *The American Printing House for the Blind* (APH: https://www.aph.org/) showed that less than 10% of children between the ages of 4 and 21 are Braille readers. This statistic is even lower for elder populations. Therefore, improving access to the Web is a priority, particularly to promote the autonomy of VIP, who do not practice Braille.

At the same time, Web information is characterized by a multi-application multi-task multi-object logic that builds complex visual structures. As such, typographical and layout properties used by web designers allow sighted users to take a large amount of information in a matter of seconds to activate appropriate reading strategies (first glance view of a web page). However, screen readers, which use voice synthesis struggle to offer equivalent non-linear reading capabilities in non-visual environments. Indeed, most accessibility software embedded in tablets and smartphones synthesize the text orally as it is overflown by the fingers. This solution is interesting but daunting when it comes to browsing new documents. In this case, the blind user must first interpret and relate snippets of speech synthesis to the organization of all page elements. Indeed, the interpretation of a web page can only be complete if the overall structure is accessible (aka. morpho-dispositional semantics). To do this, he moves his fingers over almost the entire screen to carry out a heavy and somewhat random learning phase (often too incomplete to bring out rapid reading strategies). In fact, most users rarely do so and have a "utilitarian" practice of touch devices, confining themselves to the functionalities of the web sites and interfaces they are perfectly familiar with.

To reduce the digital divide and promote the right to "stroll" to everyone, it is imperative to allow the non-visual understanding of the informational and organizational structures of web pages. When a sighted person reads a document silently, we often observe a sequence of specific and largely automated mental micro-processes: (1) the reader takes information from a first glance of all or part of the web page by an instantaneous overview (skimming); (2) he can also activate fast scans of the medium (scanning) consisting of searching for specific information within the web page. These two processes, which are more or less conscious, alternate local and global perception, and can be repeated in different combinations until individual objectives are met. Then, they can be anchored in high-level reading strategies, depending on reading constraints and intentions. As such, layout and typography play a decisive role in the success and efficiency of these processes. However, their restitution is almost absent in existing screen reader solutions.

The purpose of our research is to provide access to the visual structure of a web page in a non-visual environment, so that the morpho-dispositional semantics can be accessed by VIP and consequently enable the complete access to the informative message conveyed in a web page. For that purpose, we propose to develop a vibrotactile device (called TactiNET), which converts the visual structure of a web page into tactile landscapes that can be explored on any mobile touchscreen device. The light contrasts overflown by the fingers are dynamically captured, sent to a micro-controller, translated into vibrating patterns that may vary in intensity, frequency and temperature, and then reproduced by our actuators on the skin at the location defined by the user. In this paper, we particularly focus on the skimming strategy and leave for further work the study of scanning procedures. Consequently, we specifically tackle two research questions that are enunciated below:

- **Question 1:** What are the frequency and intensity thresholds of the device such that maximum perception capabilities can be obtained (the study of temperature is out of the scope of this paper)?
- **Question 2:** How efficient is the device to provide access and qualitative understanding of the visual shapes displayed in a web page?

The paper is organized in 6 sections. In the next section, we detail the related work. In Section 3, we provide all the technical issues of the TactiNET. In Section 4, we present the results of the study of the sensory capabilities of the device. In Section 5, we describe the experiments conducted to evaluate the efficiency of the device to reproduce the visual structure of a web page. Finally, in Section 6, we provide some discussions and draw our conclusions.

2. Related Work

Numerous devices have been proposed to attach vibrotactile actuators to the users' body to increase the perception and memorization of information. The idea of a dynamic sensory substitution system can be found as early as the 1920s as mentioned in [1]. In the specific case of the transposition of visual information into the form of tactile stimuli, a series of remarkable early experiments are described in [2], which coined the term Tactile Vision Sensory Substitution (TVSS) for this purpose. In this case, 400 solenoid actuators are integrated into the backrest of a chair and the user seated on it manipulates a camera to scan objects placed in front of him. The images captured are then translated into vibratory stimuli that are dynamically transmitted to the actuators. The spectacular results from these experiments demonstrate the power of human brain plasticity to (1) naturally exploit visual information encoded in a substitute sensory modality, (2) externalize its sensations and (3) create meaning in a manner comparable to that which would have been produced by visual perception.

A few years later, the Optacon has been proposed to offer vibrotactile feedback [3]. This particularly innovative device is capable of transposing printed texts into vibrotactile stimuli. An optical stylus and a ruler make it possible to follow the lines of a text and to reproduce the shapes of the letters dynamically under the pulp of a finger positioned on vibrotactile pins. Three weeks of learning on average were sufficient for a reading performance of about 16 words per minute, which stands for a remarkable result as it opened up the new possibility of accessing non-specific paper books, i.e., designed for sighted people. The interesting idea is to support sensory substitution through active exploration and analog rather than symbolic transposition (compared to Braille, for example in the D.E.L.T.A system [4]). Unfortunately, despite its very good reception at the time by the blind population, the marketing of the product was short due to the lack of a sustainable economic balance.

More recently, with the advent of new portable technologies and the constant increase in the power of embedded applications and actuators, we can observe many interesting studies for the use of touch in new interactions. [5] brings interesting knowledge about the potential of rich tactile notifications on smartphones with different locations for actuators, and intensity and temporal variations of vibration. Ref. [6] present a simple and inexpensive device that incorporates dynamic and interactive haptics into tabletop interaction. In particular, they focus on how a person may interact with the friction, height, texture and malleability of digital objects. Other research devices exploiting the ideas developed in TVSS are dedicated to improving the perception of blind people to facilitate their independent movements. To this end, Ref [7] propose to automatically extract the contours of images captured in the blind person's environment. This piece of information is then transposed into vibratory stimuli by a set of 48 motors integrated inside a vest worn by the subject. As such, this navigation system (Tyflos) integrates a 2D vibration array, which offers to the blind user a sensation of the 3D surrounding space.

In a task-oriented approach, these proposals do not adequately cover our needs regarding the consideration of typography and layout for the non-visual reading of web pages. Several more specific studies come closer to our perspectives. Many studies have focused on the use of textures to produce different tactile sensations during the spatial exploration of graphics [8]. This research has led to recommendations on texture composition parameters in terms of elementary shape, size, density, spacing, combination or orientation. From there, devices have been developed to facilitate access to diagrams by blind people [9–14]. However, they do not tackle the overall complexity of multi-modal web documents that may gather textual, visual, layout information, to name but a few.

To fulfill this gap, the first tactile web browser adapted to hypertext documents was proposed by [15]. Filters were applied to images and graphics to reduce the density of visual information and extract important information to be displayed on a dedicated touch screen. The text was rendered on the same touch screen in 8-dot Braille coding. This browser illustrates three main limitations we wish to remove. First, it only considers part of the layout information (layout of the elements, graphic/image/text differences), which are not sufficient to exploit the richness of the typographical phenomena (e.g., colors, weights, font size) and the luminous contrasts they induce. Second, the browser uses Braille to render text on the screen, whereas only a minority of blind people can read it and the Braille language is limited to translate typographic and layout information. Third, the user's autonomy is reduced in the choice of accessible information since the browser unilaterally decides which information is likely to be of interest. Another interesting browser has been proposed called CSurf [16] but it also relies on data filtering and the valuable information is selected by the browser itself. TactoWeb [17] is a multi-modal web browser that allows visually impaired people to navigate through the space of web pages thanks to tactile and audio feedback. It relies on a lateral device to provide tactile feedback consisting of a cell that stimulates the fingertip by stretching and contracting the skin laterally. Although it preserves the positions and dimensions of the HTML elements, TactoWeb sorts and adapts the information based on the analysis of the structure of the Document Object Model (DOM) tree. Closer to the idea of the Tactos system [9] applied to web pages, the browser proposed by [18] requires a tactile mouse to communicate the presence of HTML patterns hovering over the cursor. The mouse is equipped with two touch cells positioned on top. During the system evaluations, web pages were presented to the participants and then explored using the device. Each blind user was asked to describe the layout of the visited pages. The results indicate that while the overall layout seemed to be perceived, the description still revealed some inconsistencies in the relationship between the elements and in the perception of the size of the objects. The idea of an analogical tactile perception of a web page is appealing but only if the tactile vocabulary of the device is rich enough to transpose the visual semantics of web pages. Moreover, the disadvantage of requiring a specific web browser partially breaks the principle, which we call "design for more", on which we wish to base our solution. This approach comes from the observation that one of the reasons that weakens the appropriation of a device by a blind person is its destructive aspect. We make the hypothesis that a tool, even if it offers new useful features, may not be accepted if it prevents the use of widely tested features. The system should be able to be added to and combined with the tools classically used by a given individual, whether with a speech synthesis, a Braille track or a specific browser.

According to us, a triple objective must be guaranteed to develop successful TVSS devices in the context of non-visual access of web information: perception, action and emotion. Indeed, the typographical and layout choices contribute to giving texture and color to the perception of the document. As such, the author transmits a certain emotional value using these contrasts. Therefore, we make the hypothesis that the improvement of screen readers goes through the development of devices that make it possible to perceive the coherence of the visual structure of documents, as much as for the information it contains, and the interactions it suggests, and the emotional value it conveys.

Another important aspect for the success of powerful screen readers consists of guaranteeing conditions of appropriation. First, the system must not hinder exploratory movements, be autonomous, robust and light. It must also be discrete and inexpensive as well as be easy to remove. The device must also meet real user needs. Indeed, the multiplication of digital reading devices greatly complicates the visual structure of digital documents in general and web pages in particular. There is therefore a great need for this population to facilitate non-visual navigation on the Internet. We designed the TactiNET device, presented in the following subsection, to meet this demand.

3. TactiNET Hardware and Framework

The state of the art shows that interfaces for non-visual and non-linear access to web pages are still limited, especially with nomadic media, i.e., the blind user perceives the document only through fragments ordered in the temporal dimension alone. It is imperative to allow a non-visual apprehension of the documents that is both global and naturally interactive by giving blind people a tactile representation of the visual structures. Our ambition is to replace one's capacity for visual exploration, which relies on the luminous vibration of the screen, by a capacity for manual exploration, which relies on the tactile perception of vibrotactile or thermal (or both) actuators.

The device should translate information as faithfully as possible while preserving both informational and cognitive efficiency. The user will then be in charge of interpreting the perceived elements and will be able to make decisions that will facilitate active discovery, learning and even bypassing the initially intended uses (concept known as enactivism [19]). This idea runs counter to most of the current attempts to produce intelligent technologies by building complex applications whose Man-Machine coupling is thought upstream. In this objective, a metaphor, known as "white cane metaphor", guides both our hardware and software design, i.e., the blind person explores the world by navigating thanks to the contacts of his cane with the obstacles and materials around him. We hope that the semantics of the text visual structures will play this role for the digital exploration of documents, by creating a sensory environment made of "text sidewalks", graphical textures and naturally signposted paths orienting the movements of this "finger-cane". The TactiNET hardware [20] has been developed to offer both versatility and easy setup in the design of experiments. As shown in Figure 1, it consists of:

- A control tablet (item **1**) where all the experimental parameters can be managed and programmed (e.g., shape of the patterns, vibrations frequencies, etc.) and sent via Bluetooth to the user's tablet (item **2**).
- A user tablet (item **2**), where web pages are processed and displayed according to the graphical language. The user's five fingers coordinates are sent to the host system via a connection provided by a ZigBee dongle (item **3**).

- The host system (item **4**) where actuators can be connected through satellites. Each actuator can be either a piezoelectric vibrator or a Peltier device to provide haptic or thermal user feedback (item **5**).

Depending on the information flown over by the fingers, data packets containing the control information are transmitted from the user tablet to the host system, which can then control the piezoelectric vibrators and Peltier based on the requested amplitude, frequency and temperature values. All this hardware setup has been designed and realized at the GREYC UMR 6072 (http://www.greyc.fr) laboratory with plastic cases built with a 3D printer. A total control on all the experimental parameters is thus achieved and future extensions can be easily implemented. The host system is battery powered with a built-in USB battery management system. It is thus portable and provides more than 2 h of experiment time with full charge.

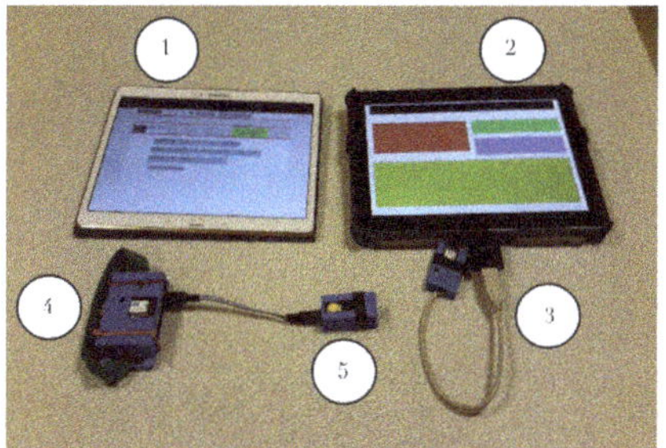

Figure 1. TactiNET: hardware setup.

3.1. Host System and Actuators Descriptions

The host system is built based on an Atmel ATZ-24-A2R processor that communicates with the user's tablet with its built-in ZigBee dedicated circuits. The host system consists of one main board with processor, battery management and communication circuits, and up to 8 daughter boards that can be stacked. Each daughter board can control up to 4 independent actuators connected via standard 3.5 mm 4 points connectors using an I2C protocol (see the satellite in Figure 2). Two kinds of actuators have been developed to provide both haptic and thermal feedback: piezoelectric and Peltier.

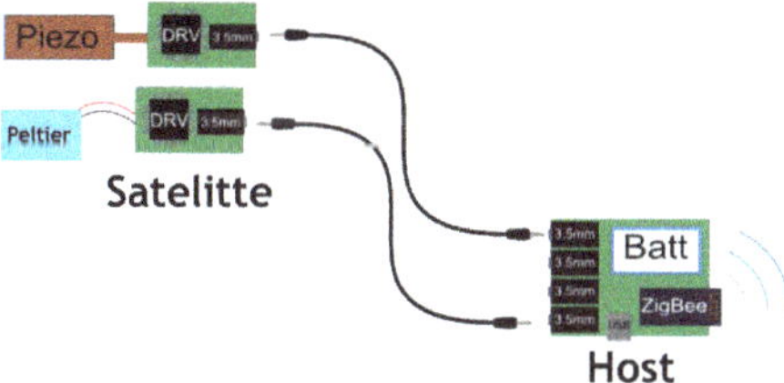

Figure 2. The host system communicates with the user's tablet via ZigBee. Satellites are connected via 3.5 mm jack connectors and contain both piezoelectric or Peltier actuators.

First, classical haptic devices usually use unbalanced mass small electric motors. In such devices, the intensity and the frequency of the vibration are completely correlated. To avoid this correlation and to have a perfect control on these two important parameters for haptic feedback, a more sophisticated actuator based on the piezoelectric effect has been used. In particular, a specific circuit from Texas Instrument was used to generate the high voltage (up to 200 V) needed to control the actuator.

Second, a simple thermal effect can be achieved using a power resistor. In such a device, only temperature increase can be generated. To have a better control on the thermal feedback, a Peltier module has been used. As such, by controlling the sign of the DC voltage, the user's skin can be cooled or heated. The DC voltage is generated with a dedicated H bridge circuit.

Each satellite board contains the actuators and the dedicated circuits needed to control it. The type of each satellite can be identified by the host system. This hardware architecture provides versatility and is easy to use. As an example, Figure 3 shows a configuration with the host system wrapped around the wrist with three satellites providing two kinds of haptic feedbacks and a thermal one.

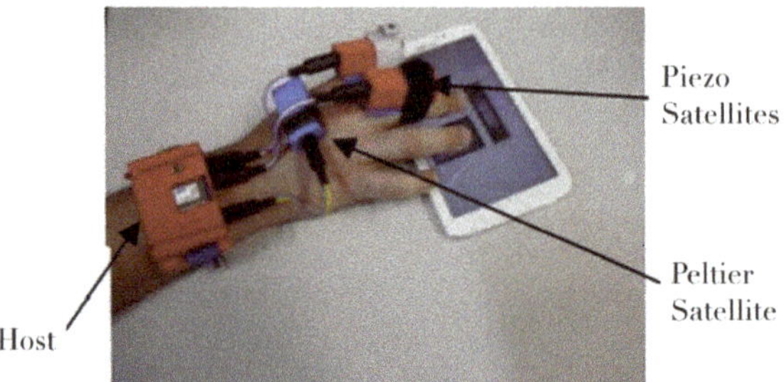

Figure 3. Example of assembly on active hand with one base, two piezoelectric and one Peltier.

3.2. Performances

The haptic feedback consists of vibrations in a frequency range from 50 Hz up to 550 Hz with a resolution of 7 Hz. The intensity of the vibration should be given in terms of skin pressure. Unfortunately, one should know the mechanical force applied by the actuators on the skin. This parameter is very difficult to obtain since it depends on the skin resistance that varies a lot from user to user and with the environment (temperature, relative humidity, etc.). As we will see in the next section, the vibration intensities are thus given by the 8 bit number used in our protocol to control the vibration (0 = no vibration and 255 = full vibration).

The thermal feedback consists of temperature variations limited to $+/-5\ ^{\circ}C$: the main limitation is due to the DC current needed that may reduce the experiment time. Up to $+/-10\ ^{\circ}C$ can be easily achieved but with serious time limitations.

With the ZigBee protocol, up to 10 host devices can be addressed. In addition, each host device can receive 8 daughter boards that can control 4 actuators. Finally, up to 320 actuators can be controlled within the TactiNET.

In summary, the device is designed as a modular experimental prototype intended for researchers who are not experts in electronics. As such, depending on the objective of the study, different combinations can easily be composed and evaluated, both in terms of the number of actuators and their type. In addition, the actuators can freely be exchanged in a plug-and-play mode and are integrated in a plastic housing made by a 3D printer. Each element has a hooking system so that it can be positioned on different parts of the user's body. In our experiments, described in the following subsections, a single satellite is used with haptic feedback and placed on the non-navigating hand.

4. The Graphical Language of the TactiNET

In this section, we first propose an empirical validation of the TactiNET framework to recognize simple shapes that simulate web page layouts, and based on demonstrated limitations, we define the foundations of a graphical language based on patterns that vary in intensity and frequency.

4.1. Towards a Graphical Language for the Tactile Output

A first experiment has been carried out in [21] that aims at pre-testing the ability of users to recognize shapes on a handheld device with the TactiNET in its minimal configuration within a non-visual environment. This configuration consists of:

- A single satellite positioned on the non-active hand: one vibrotactile actuator,
- A single dimension of variation: the brighter (respectively darker) the pixel overflown by the index finger, the lower (respectively higher) the vibration amplitude.

The experiment has been conducted with 15 sighted users (eyes closed) and 5 blind people (see Table 1) to explore, count, recognize and manually redraw different simulated web pages configurations. The conclusions of this experiment were as follows. First, the ability to distinguish the size of the shapes and their spatial relationships was assessed as highly variable in terms of exploration time (7 to 20 min in total to explore 4 configurations). The quality of manual drawings also varied from very bad to almost perfect depending on user characteristics (age, early blindness, familiarity with tactile technologies).

Table 1. Characteristics of the blind population.

User id	1	2	3	4	5
Age	63	67	59	56	36
Sex	Male	Female	Male	Female	Female
Age of blindness	0	32	25	10	15
Operating system	Linux	Linux	Windows	-	Windows
Dedicated technology	ORCA	NVDA, ORCA	JAWS, NVDA	-	JAWS

However, as demonstrated in Figure 4, the best productions are qualitatively interesting despite the relatively simple configuration of the TactiNET. Other interesting side results are worth noting. First, an encouraging learning effect was clearly demonstrated when the experience lasted no more than an hour. Second, we have identified a metric to measure a user's interest in the information overflown: the greater the interest, the proportionally greater the pressure exerted on the screen (this feature needs to be studied more deeply in future work, in particular for scanning purposes).

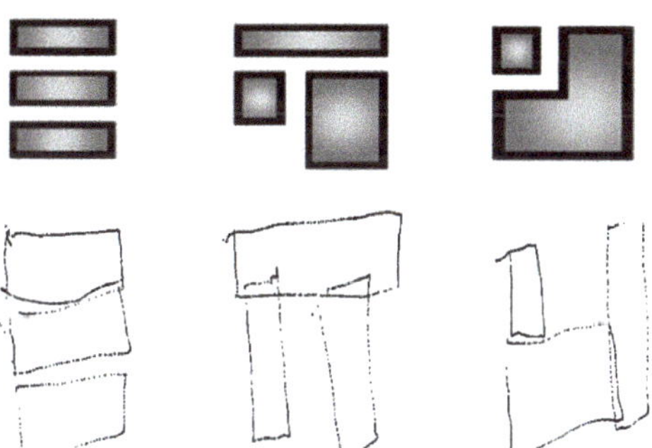

Figure 4. Original shapes and drawings of the perceived visual structures by the user.

To propose a more complete graphical language capable of handling more relationships between (1) the layout and typographic clues, and (2) the graphical patterns variable in shapes, sizes, surface and border textures, and distances, we proposed to improve the perceptive capacities of the TactiNET by optimizing the combination of different tactile stimuli, namely amplitude and frequency.

4.2. Minimum Perception Thresholds of Frequency and Amplitude

From this perspective, we designed a second experiment [22] designed to select the most perceptible frequency range of the TactiNET device. Each frequency value studied was further combined with an amplitude value either constant or increased by a slight random variability in order to incorporate a texture effect that may enable more perceptual capacities. For that purpose, we conducted an experiment that crossed 3 groups of users (38 sighted children, 25 sighted adults and 5 blind adults, the same population as in Section 4.1) and consisted of a series of comparison tests on a touch screen divided into two distinct areas. The user had to decide whether the stimuli perceived when flying over each zone differed. All tests were performed with a single actuator. The user had to explore the tablet with the finger of one hand, and with the actuator placed on the other hand. The only question asked after each exploration with no time limit was whether the two stimuli on the left and right of the interface were identical. Initially, all amplitude values were set to a fixed value, and only the frequency values changed from one stimulus to the other. To minimize interference and maximize the fluidity of the experiment, a second tablet dedicated to the experimenter was connected to the first one by a Bluetooth connection. It was equipped with an interface to quickly and remotely control the presentation and the successive value of the series of stimuli. Each series consisted of a fixed reference value sent to the left side of the tablet and a variable comparison value sent to the right side.

The main result indicates that participants were collectively more sensitive to differences of frequencies close to 300 Hz. This perceptive ability becomes individualized and deteriorates as one moves away from this value. This result can be positively compared to [23], which indicates that tactile navigation on digital devices is more sensitive to vibration with frequencies within the range of 200–250 Hz. On the other hand, the experiment did not show any significant effect of amplitude noise, age or blindness (except for children who are significantly more sensitive to ascending than descending series). To summarize, the conclusions concerning the first grammatical rules applicable to our graphical language are declined according to the 5 frequency ranges listed below:

- Between 50 Hz and 150 Hz, the minimum perceptual threshold is 15 Hz.
- Between 150 Hz and 250 Hz, the minimum perceptual threshold is 13 Hz.
- Between 250 Hz and 350 Hz, the minimum perceptual threshold is 7 Hz.
- Between 350 Hz and 450 Hz, the minimum perceptual threshold is 10 Hz.
- Between 450 Hz and 550 Hz, the minimum perceptual threshold is 15 Hz.

These frequency values must be combined with amplitude values to increase the expressive power of the graphical language. Consequently, we designed a third experiment [24] to select the most perceptible amplitude range. Each studied amplitude value was further combined with the optimized frequency value obtained in the previous experiment. The protocol was very similar to the previous one excepted for the population (20 sighted adults and 5 blind adults (the same population as in Section 4.1), the reference values for amplitude ranged from 55 to 255 (Set to this range to be included in the communication frame and then translated by the micro-controller in amplitude value.), and the "staircase" and "up-and-down" method [25] used to determinate the amplitude perceptive threshold.

Many amplitude values were compared to three reference amplitudes by crossing the two populations of users. The best perceptive results are in an interval around the value 55 regardless of the experimental condition. Nevertheless, although measuring less sensitive perceptual thresholds, the higher values reveal a significant difference between the sighted and blind groups. Based on these results, we increased our grammar by the following 3 rules:

- Between 0 and 100, the minimum perceptual threshold is 12.
- Between 100 and 200, the minimum perceptual threshold is 48.
- Between 200 and 255, the minimum perceptual threshold is 45.

Through these experiments, we began to weave links between a graphical language and basic parameters of vibrotactile stimuli. In our last experiment, which is presented in the following section, we attempt to relate these results to the visual structures extracted from a corpus of 900 web pages.

5. Recognition of Web Page Structures

To test the perceptual capacities of the TactiNET through its graphical language, we propose an exploratory experiment based on the semi-automatic conversion of the global visual structures of web pages into vibrotactile stimuli. This conversion is achieved by dividing a given web page into meaningful clusters of bounding-boxes represented by rectangular shapes, and by translating them into vibrotactile stimuli. The final objective is to evaluate how these macro-structures can be perceived in a non-visual environment by using the graphical language of the TactiNET.

5.1. Protocol Design

To evaluate the perceptive capacities of our device, we propose to run an exploratory experiment, which tackles a small number of web pages from three different domains: tourism, e-commerce and news (these are the three domains that are tackled by the academic-industrial consortium of the TagThunder project funded by BPI France).

In particular, three representative web pages have been automatically chosen from a corpus of 900 web pages equally distributed by domain. The representative web page of a given category is the one that obtains the highest similarity score when compared to all the other web pages in the same set according to a similarity function proposed in [26]. The measure of similarity is based on the number of intersection points between the rectangular bounding-boxes that make up two pages when they are superimposed. Therefore, the web page that shows the maximum intersection points with all the web pages of the category is chosen as the most representative of the category. Based on this measure, the three representative web pages that have emerged from each category are www.francetourisme.fr (tourism), www.asdiscount.com (e-commerce), and www.fdlm.org (news) (see Figure 5).

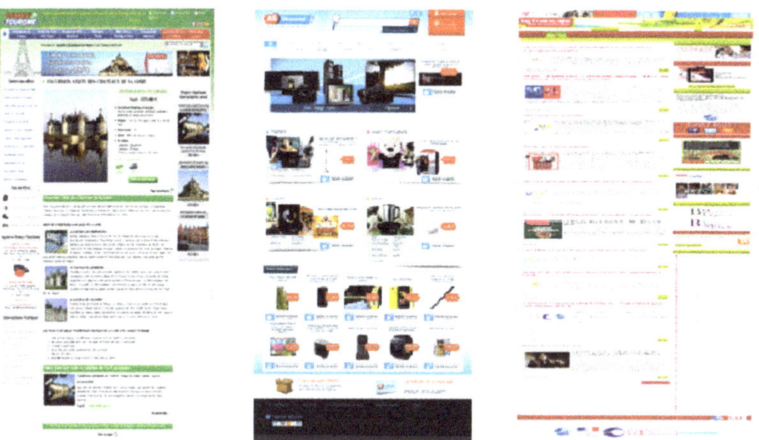

Figure 5. Three representative web pages: tourism (**left**), e-commerce (**middle**), and news (**right**).

After choosing the representative web pages, an agglomerative graph-based clustering algorithm described in [27] has been applied on each of them to obtain meaningful clusters representing their visual macro-structures. The algorithm starts with a web page and a given number of clusters to discover. Then, three phases are processed: (1) extracting the visual parameters of HTML elements (vision-based phase), (2) filtering and reorganizing HTML elements (DOM-based phase), and (3) clustering the extracted bounding-boxes (agglomerative graph-based phase).

For this experiment, we fixed the number of zones to 5. This number comes from [28], who showed that there is a limit in terms of the amount of information a person can receive, process, and remember. The theory, known as Miller's law, proposes that the average number of objects one can hold in working memory is 7 +/− 2. Therefore, to guarantee that the number of chosen clusters does not affect the performance of the participants, we chose 5 clusters as the minimum information to be handled. This process is illustrated in Figure 6, where the rectangular shapes result from the agglomerative graph-based clustering algorithm applied to each representative web page. Please note that the clustering shows differences in surface, location, orientation, and spatial relations. The results from the clustering have then been adapted to our experimental setup to take into account the recommendations of [12] in terms of rendering forms for non-visual navigation on touchscreen devices. Figure 7 shows the adapted presentations of the vibrating pages representing each page of each category (the colors have no other interest than to clearly distinguish the areas for the experimenter). After adapting each representative web page, a particular vibrotactile feedback should be dedicated to each cluster. To exploit the light contrasts produced by the tablet, the relationship between the visual structure of the zones and the calculated vibrotactile stimulus is based on the standard deviation of the gray values of the pixels. The greater the visual contrast, the higher the standard variation value, the greater the strength of the tactile stimulus should be. Therefore, based on the results presented in the previous section in terms of perception thresholds, the standard deviation value calculated for each zone is proportionally associated with an amplitude value as mentioned in Table 2, while a constant value of 304.6875 Hz is assigned to frequency. A basic vibrating page was also constructed to serve as a reference by setting the amplitude to the optimum value of 55 for all zones.

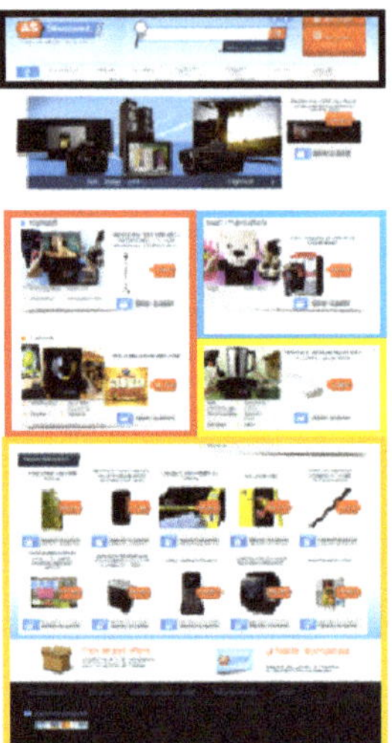

Figure 6. Clustering process of representative web pages.

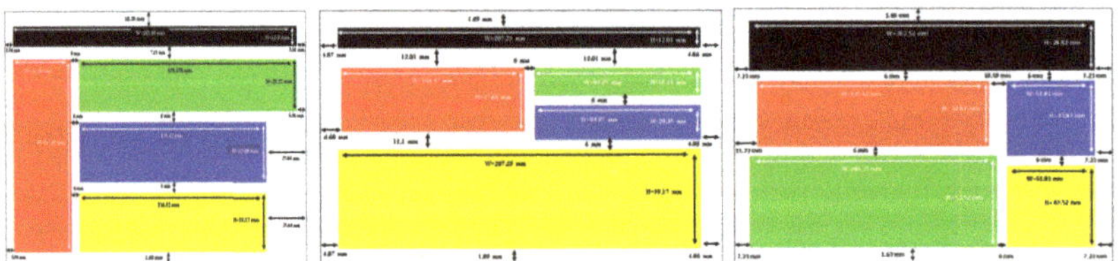

Figure 7. Adapted segmentation for experimental purposes.

Table 2. Amplitude values for the vibrotactile stimuli.

Web Page	Black Zone	Red Zone	Green Zone	Blue Zone	Yellow Zone
Tourism	69	50	78	73	75
Commerce	79	80	72	65	95
News	69	57	52	77	31

By optimizing both the vibration feedback parameters and the difference in the generated visual structures, the experiment consists of making the hypotheses that (1) the TactiNET should increase the ability to discern different categories of vibrating pages (i.e., whether two pages shown are similar or different) and (2) better performance in this task should be achieved when using values of varying amplitudes. The experimental setup is illustrated in Figure 8. Eleven users participated in this experiment (5 sighted and 6 blind) as demonstrated in Table 3. The group of the 6 blind participants is balanced in terms of gender and the precocity of their blindness and their average age is around 54 years old. Five of the blind population regularly use personal computers but in different proportions (from 1 to 8 hours a day), and only one of them uses a smartphone to surf on the web. In parallel, all sighted participants are women with an average age of around 26 years old.

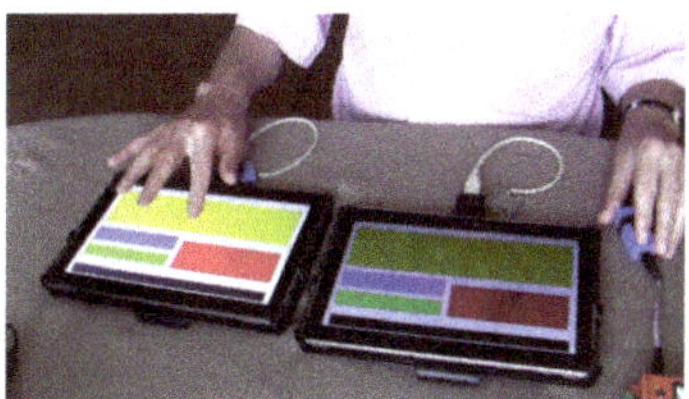

Figure 8. Comparing web page structures depending on associated amplitude values.

Table 3. Characteristics of the blind and sighted population.

User Id	Vision Status	Age	Sex	Age of Blindness
D1	blind	55	Female	0
D2	blind	66	Male	0
D3	blind	39	Female	8
D4	blind	40	Male	0
D5	blind	69	Female	32
D6	blind	60	Male	27
V1	sighted	27	Female	-
V2	sighted	26	Female	-
V3	sighted	25	Female	-
V4	sighted	29	Female	-
V5	sighted	25	Female	-

Overall, 36 web page structures comparisons have been performed by each participant. There are 3 comparisons of identical structures and 3 comparisons with different page structures. All comparisons are repeated 3 times to avoid random selection. In addition, all comparisons are subject to 2 amplitude conditions: variable or fixed. Under the same conditions as the previous experiment, each task included a series of tests, which present the compared structures on two touchscreen tablets. After the exploration phase, the participant had to decide whether both vibrotactile structures were identical or not. Finally, at the end of all 36 comparisons, the user had to draw the last discovered structure on a A4 sheet of paper.

5.2. Analysis of the Results

We are aware that the number of blind participants is not sufficient to perform statistically significant tests. However, gathering a large population of visually impaired people is (fortunately) not an easy task. Consequently, the analysis of this experiment will be exploratory and preparatory to determine specific hypotheses and to validate identified trends to be developed in a future larger protocol. Figure 9 gathers results for all the comparisons of all pairs of vibrotactile pages, representative of the following categories: news (N), e-commerce (E), and tourism (T). The scores are distributed in the four tables according to the 6 blind participants (D1 to D6), the 5 sighted participants (V1 to V5), and the condition of the amplitude value (set to 55 or associated with the contrast value of the original area set in Table 2).

COMPARISONS OF BLIND PEOPLE	Fixed Amplitude 55								Variable Amplitude							
	Similar Structures			*Dissimilar Structures*			Correct Answers by Subject		*Similar Structures*			*Dissimilar Structures*			Correct Answers by Subject	
Subject ID	*N↔N*	*E↔E*	*T↔T*	*N↔E*	*N↔T*	*E↔T*			*N↔N*	*E↔E*	*T↔T*	*N↔E*	*N↔T*	*E↔T*		
D1	0	0	0	3	3	3	9	50.0%	0	0	2	3	3	3	11	61.1%
D2	0	1	2	2	2	3	10	55.6%	0	2	3	2	3	2	12	66.7%
D3	3	3	2	2	3	2	16	88.9%	3	2	3	2	3	3	16	88.9%
D4	2	1	3	3	3	2	14	77.8%	3	2	3	2	2	2	14	77.8%
D5	1	1	0	1	2	3	8	44.4%	1	0	0	3	3	3	10	55.6%
D6	1	0	1	2	3	2	9	50.0%	2	1	0	2	1	1	7	38.9%
Correct Answers by Pair of Pages	7	6	9	13	16	15	Total Correct Answers		9	7	11	14	15	14	Total Correct Answers	
	38.9%	33.3%	50.0%	72.2%	88.9%	89.3%			50.0%	38.9%	61.1%	77.8%	83.3%	77.8%		
Correct Answers by Similarity of Structures	22			44			$\frac{66}{108} = 61.1\%$		27			43			$\frac{70}{108} = 64.8\%$	
	40.7%			81.5%					50%			79.6%				

COMPARISONS OF SIGHTED PEOPLE	Fixed Amplitude 55								Variable Amplitude							
	Similar Structures			*Dissimilar Structures*			Correct Answers by Subject		*Similar Structures*			*Dissimilar Structures*			Correct Answers by Subject	
Subject ID	*N↔N*	*E↔E*	*T↔T*	*N↔E*	*N↔T*	*E↔T*			*N↔N*	*E↔E*	*T↔T*	*N↔E*	*N↔T*	*E↔T*		
V1	0	2	0	2	1	2	7	38.9%	3	1	2	3	2	3	14	77.8%
V2	1	0	0	3	3	3	10	55.6%	1	1	0	3	3	3	11	66.1%
V3	3	1	2	3	3	3	15	83.3%	1	2	2	3	3	3	14	77.8%
V4	0	3	1	3	3	2	12	66.7%	0	1	2	1	1	2	7	38.9%
V5	1	2	1	3	2	2	11	61.1%	1	0	0	3	2	3	9	50.0%
Correct Answers by Pair of Pages	5	8	4	14	12	12	Total Correct Answers		6	5	6	13	11	14	Total Correct Answers	
	33.2%	53.3%	26.7%	93.3%	80.0%	80.0%			40.0%	33.3%	40.0%	86.7%	73.3%	93.3%		
Correct Answers by Similarity of Structures	17			38			$\frac{55}{90} = 61.1\%$		17			38			$\frac{55}{90} = 61.1\%$	
	37.8%			84.4%					37.8%			84.4%				

Figure 9. Results of all comparisons.

Several general observations can be specified. First, the results are globally homogeneous between the two populations regardless of the amplitude condition with a ratio of about 60% correct answers overall (exactly between 61.1% and 64.8%). This score seems promising as (1) 5 out of 6 blind people never use tactile devices, (2) none of them had any training time, and the experience was long (on average 1.5 h) and boring, and (4) the device was in its minimal configuration i.e., one single vibrator and a maximum of one dimension of vibratory variation. However, we note a slight superiority of almost 4% for the blind population to correctly classify pairs with the variable amplitude. This remark is important as this superiority comes exclusively from the case of comparisons between similar structures. While the results in the four tables show that it is (logically) twice as easy to identify a dissimilarity between two vibrotactile structures as it is to be certain of their identity, the blind seem more able (by 10%) to exploit the variation in amplitude to remove their doubts about the similarity of two structures.

5.2.1. Redrawing Task

Some remarks concern the quality of the drawings produced at the end of the experiment, as no prior information was given on the nature of the elements explored. The sighted people (eyes closed) were more comfortable in representing their perception naturally in the form of rectangles, but the drawing was often more complex than the structure explored (see Figure 10). In parallel, three of the non-sighted people could not abstract this notion and reproduced the path followed by their finger in the form of lines (see Figure 11).

However, two blind people reproduced the structures very faithfully, even though no formal indication had been given on the vibratory stimulus support (see Figures 12 and 13). In any case, this exercise seemed to be an interesting task, even with a blind population of users, to evaluate the capacity of our stimuli to construct a mental representation of the visual structure. With a statistically more appropriate number of users, Figure 14 might show a correlation between the scores given to the drawings by experts and the comparison ones. This may be particularly true when extracting exclusively results from the blind subjects (Figure 14 right). In this case, the correlation appears to be more strongly linear than for the overall observation (Figure 14 left), where the V3 data may be an outlier.

Figure 10. Sample drawings of V1 (middle) and V3 (right) in reference to the news structure.

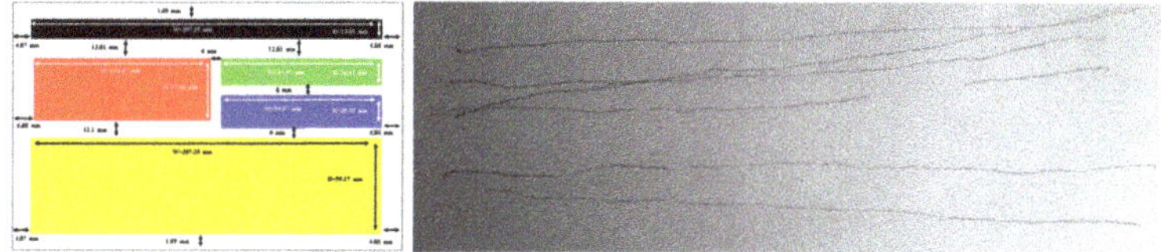

Figure 11. Sample Drawings of D4 in reference to the e-commerce structure.

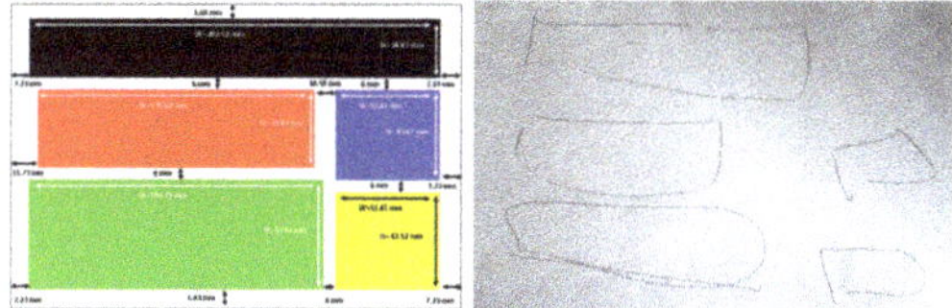

Figure 12. Sample drawings of D2 in reference to the news structure.

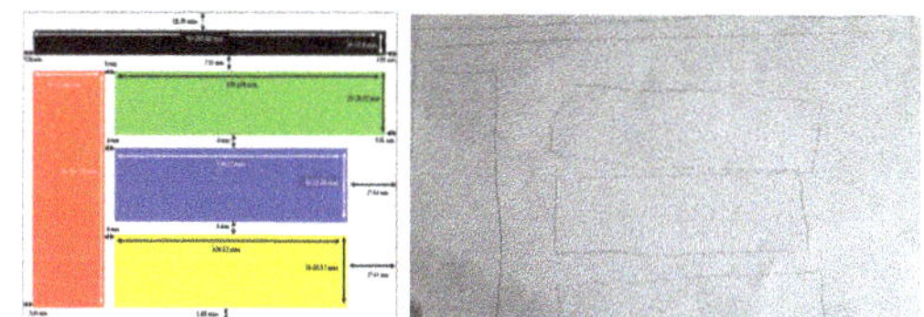

Figure 13. Sample drawings of D3 in reference to the tourism structure.

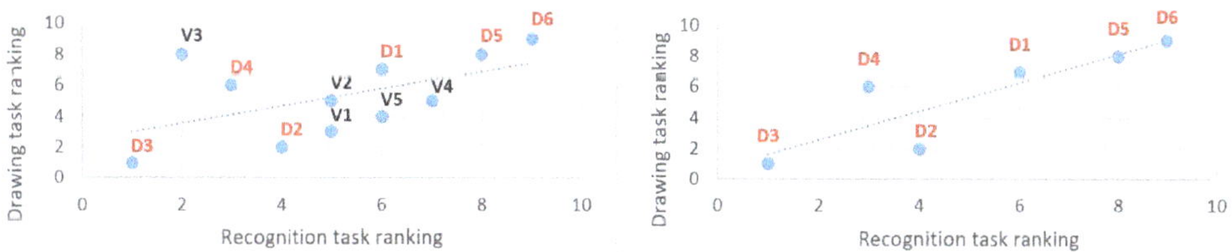

Figure 14. Recognition and drawing tasks: correlation for all subjects (**left**) and for the blind (**right**).

5.2.2. Navigation Strategies

As all the experiments were filmed, we were able to observe the different strategies used by the participants to explore the tablet touchscreen. A first typology is provided in Figure 15. What seems remarkable to us is the richness of the micro-strategies used and therefore the large number of possible combinations to anchor them in a global strategy (see Table 4). This experience does not yet allow us to define a clear relationship between combinations of micro-strategies and overall efficiency for the recognition of visual text structures. Nevertheless, the relationship with the pre-perceptual visual abilities seems obvious enough to extend this line of research, i.e., the aim is to lay a solid foundation for the exploitation of non-visual scanning processes and thus the development of effective tactile reading strategies. A preliminary analysis can be produced from the following three observations:

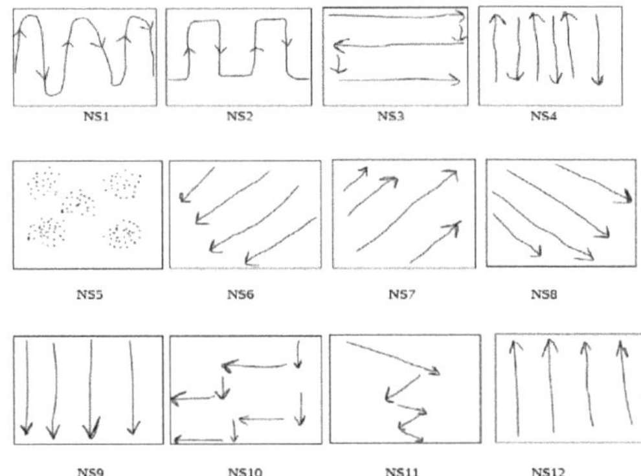

Figure 15. Navigation micro-strategies (NS) of blind and sighted participants.

Table 4. Navigation strategies (NS) chosen by each participant.

Subject ID	Chosen Navigation Strategies
D1	NS1, NS2, NS3, NS4, NS5, NS12
D2	NS3, NS4, NS12
D3	NS1, NS3, NS4, NS12
D4	NS3, NS4, NS10, NS12
D5	NS3, NS8, NS9
D6	NS3, NS4, NS8, NS9
V1	NS3, NS4, NS6, NS7, NS11, NS12
V2	NS3, NS4, NS12
V3	NS3, NS4
V4	NS3, NS4, NS12
V5	NS3, NS4

- The majority of participants start navigating the tablets from the left to right. This might be due to the cultural habits related to the writing and reading directions.
- The participants use between two to six micro-strategies, which can be arranged into three classes: (1) continuous navigation taking information in both horizontal and vertical dimensions (NS1, NS2, NS3, NS10, NS11), (2) navigation using only one direction (horizontal, vertical or diagonal - NS4, NS6, NS7, NS8, NS9, NS12) and (3) navigation using none of these possibilities (NS5).
- The micro-strategy NS3 is used by all participants as well as NS4 to the exception of D5. It is also noticeable that there were remarkable differences between the participants in the time and the speed of using each micro-strategy.

These remarks lead us to make several hypotheses. As the structure being flown over is not known in advance, the natural strategies implemented are preferentially based on a continuous horizontal and vertical course of the entire screen. However, the efficiency of information capture is degraded by the sequence of too many or too different micro-strategies. These findings would participate to explain the three lowest scores of blind people: D1 (too many micro-strategies), and D5 and D6 (use of too many dimensions). The most effective strategies (D2, D3 and D4) have in common a two-step recognition.

The information is first taken by a continuous path in both horizontal and vertical dimensions of the screen, and then followed by a verification, essentially vertical, to remove some uncertainties.

5.2.3. Other Meaningful Results

Finally, looking at the detailed results by participant or by category of visual structures, some elements of discussion also emerge. First, only the "tourism" web page structure explored by blind people seems to influence the scores, i.e., scores are higher for blind people than for sighted participants. Indeed, by looking at the left-hand bar graphs (blind) of Figures 16 and 17, the comparisons involving this visual structure are systematically (to one exception) superior in both the "similar" (yellow bar) and the "different" (brown and cyan bars) conditions. The explanation we propose is to be found in the congruence between the shapes flown over and the amplitude value chosen. Indeed, the tourism category is the only one for which shapes close in surface and size are associated with values of close intensity (see Figure 7 and Table 2). This conclusion, if proven by further experiments on a vast population, is very interesting to support our approach because it legitimizes the use of objective criteria (variance in contrast), extracted from the source document and transformed in an analogical way, to build a coherent tactile landscape. In any case, this also requires us (in the future) to weight the amplitude value associated with the change in contrast of the zone by an area value of the zone.

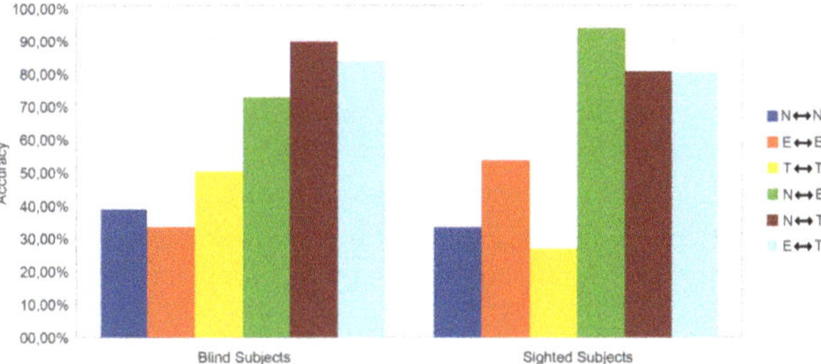

Figure 16. Accuracy by category - amplitude set at 55.

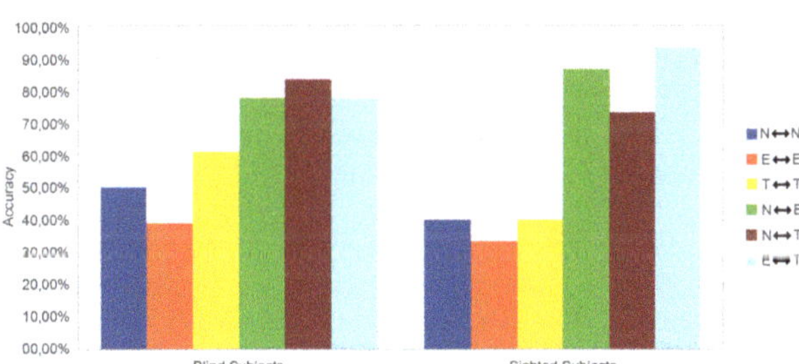

Figure 17. Accuracy by category - variable amplitude.

Second, there seems to be a tendency for scores to be higher, the earlier the onset of blindness is, as shown in Table 5. In addition to the conditions for the occurrence of blindness, we will note in our results a tendency towards an easier task for blind people who are familiar with new tactile technologies, and for those who spend a lot of time surfing the web. In fact, the best performance is attributable to the only blind woman in the protocol (i.e., D3 with nearly 90% correct answers under all conditions), who is the youngest, has an iPhone equipped with VoiceOver, and is connected for more than 10 h a day.

Table 5. Congenital vision loss effect on accuracy.

	Fixed Amplitude		Variable Amplitude	
Type of Participants	Similar Struct.	Dissimilar Struct.	Similar Struct.	Dissimilar struct.
All blind	40.7%	81.5%	50.0%	79.6%
Congenital visual loss	50.0%	86.1%	63.9%	83.3%

6. Conclusions and Perspectives

In this paper, we developed a vibrotactile framework called TactiNET for the active exploration of the layout and the typography of web pages in a non-visual environment, being the idea to access the morpho-dispositionnal semantics of the message conveyed on the web.

For that purpose, we first built an experimental device allowing the analogical transposition of the light contrasts emitted by a touch-sensitive tablet into vibratory and thermal stimuli. The main ergonomic constraints were to be easily positionable on any part of the body, modular in terms of the quality and the quantity of actuators, inexpensive (less than 100 dollars), robust and light.

We then tuned the device to lay the bricks of a tactile language based on the expressive capacity of the stimuli produced. This work was initiated by the study of the minimum thresholds of perception of the frequency and the amplitude of the vibration.

Finally, we evaluated the ability of the TactiNET to allow the correct categorization of web pages of three domains, namely tourism, e-commerce and news presented through a vibrotactile adaptation of their visual structure. Although exploratory, the experiments are particularly encouraging reinforced by the fact that we deliberately chose very unfavorable conditions, i.e., (1) heterogeneity of the blind population in terms of age, habituation to tactile technologies and web browsing, onset of blindness, and (2) minimal configuration of our device with only one vibrotactile actuator and one maximum dimension of variation. Despite this, the interesting hypotheses that we retain are:

- blind people tend to affirm the similarity between two structures better than sighted people, especially when the relationships between the form overflown and the perceived intensities are consistent;
- blind people seem to be capable of imagining the shapes they have felt without any prior indication of the stimuli;
- users seem to develop a rich set of micro-strategies to browse the vibrating touch screen;
- the regular use of touch technologies and the number of daily hours spent on the web seem to be a positive factor for the appropriation of the device.

This first exploratory study opens up many avenues of research. For example, one possible direction is to refine on pressure, spatial and temporal criteria the typology of micro-strategies for exploring forms. The objective will be to analyze the relationships between micro-strategies and macro-strategies to access to information. We believe that through this interactive alternation between local and global perceptions, our natural scanning and skimming capabilities can be exploited, whether in the visual or the tactile modality.

One of the main perspectives of this work is also to experimentally explore more complex configurations of our device and thus improve the expressiveness of our tactile language. First, we will propose to study the association of visual and thermal parameters. Second, work is in progress to evaluate the possibility of combining tactile stimuli with audio stimuli [29]. Third, increasing the number of concurrent stimuli (up to 320) by associating a vibrotactile actuator to several fingers may enable new sensory experiences as the touch screen might be accessed by a multi-sensory device.

Finally, we will note the emergence in our experience and by informal discussions with blind people of a possible extension of the language. Indeed, one of the frequent difficulties encountered by blind people is the difficulty of having an *a priori* idea of the size of a document. In our experiments, some of the shapes chosen to calibrate the device had a black to white gradation over their entire surface (i.e., the higher the gray level, the stronger the vibration of the actuator). Therefore, the gradation produced a continuously decreasing vibration as the finger flew over it. Many blind people reported feeling differences in the speed of this decrease from the very beginning of their exploration of the shape. In fact, we constructed by chance a stimulus that allowed blind people to anticipate the size of an area. Therefore, it did not seem to be necessary for blind people to fly over more than a few centimeters to discriminate these differences and to know in advance the complete travel time of the shape. Consequently, we will propose to study the interest of a triple relationship (size of a zone, surface gradient and amplitude strength).

Author Contributions: W.S., F.M. and P.B. conceived and designed the experiments; W.S. and J.-M.R. designed TactiNet hardware and software; F.M. and G.D. supervised all the process and wrote the paper. All authors have read and agreed to the published version of the manuscript.

Funding: This work has been founded from 2013 to 2016 by the French National Research Agency (ANR ART-ADN Project; GRANT NUMBER: ANR-12-SOIN-0003); and from 2018 to present by the Programme d'Investissement d'Avenir (PIA) and the French Public Investment Bank (TagThunder Project).

Conflicts of Interest: The authors declare no conflict of interest.

References

1. Choi, S.; Kuchenbecker, K.J. *Vibrotactile Display: Perception, Technology, and Applications*; IEEE: Piscataway, NJ, USA, 2013; Volume 101, pp. 2093–2104, ISSN: 0018-9219. [CrossRef]
2. Bach-y-Rita, P.; Collins, C.C.; Saunders, F.A.; White, B.; Scadden, L. Vision substitution by tactile image projection. *Nat. Int. Wkly. J. Sci.* **1969**, *221*, 963–964. [CrossRef] [PubMed]
3. Goldish, L.H.; Taylor, H.E. The Optacon: A Valuable Device for Blind Persons. In *New Outlook for the Blind*; The American Foundation for the Blind: Arlington County, VA, USA, 1974; pp. 49–56.
4. Conter, J.; Alet, S.; Puech, P.; Bruel, A. A low cost, portable, optical character reader for blind. In Proceedings of the Workshop on the Rehabilitation of the Visually Impaired, the Institute for Research on Electromagnetic Waves of the National Research Council, Florence, Italy, 30 April 1986; Springer: Berlin, Germany, 1986; Volume 47, pp. 117–125, Print ISBN 978-94-010-8402-4, Online ISBN 978-94-009-4281-3, ISSN: 0303-6405. [CrossRef]
5. Sahami, A.; Holleis, P.; Schmidt, A.; Häkkilä, J. Rich Tactile Output on Mobile Devices. In Proceedings of the Ami'08 European Conference on Ambient Intelligence, Nürnberg, Germany, 19–22 November 2008; pp. 210–221, ISBN 978-3-540-89616-6. [CrossRef]
6. Marquardt, N.; Nacenta, M.; Young, J.; Carpendale, S.; Greenberg, S.; Sharlin, E. The Haptic Tabletop Puck: Tactile Feedback for Interactive Tabletops. In Proceedings of the Interactive Tabletops and Surfaces - ITS'09, Banff, AB, Canada, 23–25 November 2009; pp. 93–100.
7. Dakopoulos, D.; Bourbakis, N. Towards a 2D tactile vocabulary for navigation of blind and visually impaired. In Proceedings of the IEEE International Conference on Systems, Man and Cybernetics, San Antonio, TX, USA, 11–14 October 2009. [CrossRef]
8. Hollins, M.; Bensmaia, S.J.; Roy, E.A. Vibrotaction and texture perception. *J. Behav. Brain Res.* **2012**, *135*, 51–56. [CrossRef]
9. Tixier, M.; Lenay, C.; Le-Bihan, G.; Gapenne, O.; Aubert, D. Designing Interactive Content with Blind Users for a Perceptual Supplementation System. In Proceedings of the 7th International Conference on Tangible, Embedded and Embodied Interaction, Barcelona, Spain, 10–13 February 2013; pp. 229–236, ISBN 978-1-4503-1898-3.[CrossRef]
10. Petit, G.; Dufresne, A.; Levesque, V.; Hayward, V.; Trudeau, N. Refreshable Tactile Graphics Applied to Schoolbook Illustrations for Students with Visual Impairment. In Proceedings of the ACM ASSETS'08, 10th International ACM SIGACCESS Conference on Computers and Accessibility, Halifax, NS, Canada, 13–15 October 2008; pp. 89–96, ISBN: 978-1-59593-976-0. [CrossRef]
11. Levesque, V.; Hayward, V. Tactile Graphics Rendering Using Three Laterotactile Drawing Primitives. In Proceedings of the 16th Symposium on Haptic Interfaces For Virtual Environment and Teleoperator Systems, Reno, NV, USA, 13–14 March 2008; pp. 429–436. [CrossRef]
12. Giudice, N.; Palani, H.; Brenner, E.; Kramer, K. Learning non-visual graphical information using a touch-based vibro-audio interface. In Proceedings of the ASSETS '12 Proceedings of the 14th International ACM SIGACCESS Conference on Computers and Accessibility, Boulder, CO, USA, 22–24 October 2012; pp. 103–110, ISBN 978-1-4503-1321-6. [CrossRef]

13. Martínez, J.; García, A.; Martínez, D.; Molina, J.; González, P. Texture Recognition: Evaluating Force, Vibrotactile and Real Feedback. In Proceedings of the 13th International Conference, Human-Computer Interaction – INTERACT'2011, Lisbon, Portugal, 5–9 September 2011; Volume 6949, pp. 612–615, Printed ISBN: 978-3-642-23767-6, Online ISBN: 978-3-642-23768-3. [CrossRef]
14. Pappas, T.; Tartter, V.; Seward, A.; Genzer, B.; Gourgey, K.; Kretzschmar, I. Perceptual Dimensions for a Dynamic Tactile Display. In Proceedings of the SPIE Society for Photo-Instrumentation Engineers, Human Vision and Electronic Imaging, San Jose, CA, USA, 19–22 January 2009; Volume 7240, ISBN 978-0-8194-7490-2.
15. Rotard, M.; Knödler, S.; Ertl, T. A Tactile Web Browser for the Visually Disabled. In Proceedings of the Sixteenth ACM Conference on Hypertext and Hypermedia, HYPERTEXT '05, Salzburg, Austria, 6–9 September 2005; pp. 15–22, ISBN 1-59593-168-6. [CrossRef]
16. Mahmud, J.; Borodin, Y.; Ramakrishnan, I. Csurf: A context-driven non-visual web-browser. In Proceedings of the WWW'07 the 16th International Conference on the World Wide Web, Track: Browsers and User Interfaces, Session: Smarter Browsing, Banff, AB, Canada, 8–12 May 2007; ISBN 978-1-59593-654-7. [CrossRef]
17. Petit, G.; Dufresne, A.; Robert, J.M. Introducing TactoWeb: A Tool to Spatially Explore Web Pages for Users with Visual Impairment. In Proceedings of the UAHCI 2011, 6th International Conference of Universal Access in Human-Computer Interaction, Design for All and eInclusion, Orlando, FL, USA, 9–14 July 2011; Volume 6765, pp. 276–284; Print ISBN 978-3-642-21671-8, Online ISBN 978-3-642-21672-5, Series ISSN: 0302-9743. [CrossRef]
18. Kuber, R.; Yu, W.; O'Modhrain, M. Tactile Web Browsing for Blind Users. In Proceedings of the Haptic Audio Interaction Design'10, Copenhagen, Denmark, 16–17 September 2010; pp. 75–84, Print ISBN 978-3-642-15840-7, Online ISBN 978-3-642-15841-4, ISSN: 0302-9743. [CrossRef]
19. Thompson, E. Chapter 1: The enactive approach. In *Mind in Life:Biology, Phenomenology, and the Sciences of Mind*; Harvard University Press: Cambridge, MA, USA, 2010. ISBN 978-0674057517.
20. Radu, C. ART-ADN: Technical Document—A technical report of TactiNET. In *Groupe de Recherche en Informatique, Image, Automatique et Instrumentation de Caen*; Project ART-ADN; Laboratory GREYC: Caen, France, 2015.
21. Safi, W.; Maurel, F.; Routoure, J.-M.; Beust, P.; Dias, G. An Empirical Study for Examining the Performance of Visually Impaired People in Recognizing Shapes through a vibro-tactile Feedback. In Proceedings of the ASSETS2015, the 17th International ACM SIGACCESS Conference on Computers and Accessibility, Lisbon, Portugal, 26–28 October 2015.
22. Safi, W.; Maurel, F.; Routoure, J.-M.; Beust, P.; Molina, M.; Sann, C. An Empirical Study for Examining the Performance of Visually Impaired People in Discriminating Ranges of Frequencies through Vibro-tactile Feedbacks. In Proceedings of the International Workshop on the Future of Books and Reading in Human-Computer Interaction Associated to NordiCHI 16, the 9th Nordic Conference on Human-Computer Interaction, Gothenburg, Sweden, 23–27 October 2016.
23. Moore, D.; Mccabe, G.; Craig, B. *Introduction to the Practice of Statistics*, 6th ed.; W. H. Freeman and Company: New York, NY, USA, 2003. ISBN-13: 978-1-4292-1623-4
24. Safi, W.; Maurel, F.; Routoure, J.-M.; Beust, P.; Molina, M.; Sann, C.; Guilbert, J. Which ranges of intensities are more perceptible for non-visual vibro-tactile navigation on touch-screen devices. In Proceedings of the VRST 2017, Gothenburg, Sweden, 8–10 November 2017.
25. Treutwein, B. Minireview: Adaptive psychophysical procedures. *Vis. Res.* **1995**, *35*, 2503–2522. [CrossRef]
26. Juan, W.; Zhang, J.; Yan, J.; Liu, W.; Guangming, S. Design of a Vibrotactile Vest for Contour Perception. *Int. J. Adv. Robot. Syst.* **2012**, *9*. [CrossRef]
27. Safi, W.; Maurel, F.; Routoure, J.-M.; Beust, Dias, G. A Hybrid Segmentation of Web Pages for Vibro-Tactile Access on Touch-Screen Devices. In Proceedings of the VL@COLING 2014, Dublin, Ireland, 23 August 2014; pp. 95–102.
28. Miller, G.A. The magical number seven, plus or minus two: Some limits on our capacity for processing information. *J. Psychol. Rev.* **1956**, *63*, 81–97. [CrossRef]
29. Maurel, F.; Dias, G.; Ferrari, S.; Andrew, J.J.; Giguet, E. Concurrent Speech Synthesis to Improve Document First Glance for the Blind. In Proceedings of the 2nd International Workshop on Human-Document Interaction (HDI 2019) in Conjunction with IAPR/IEEE ICDAR 2019, Sydney, Australia, 22–25 September 2019.

Article

Smart Tactile Sensing Systems Based on Embedded CNN Implementations

Mohamad Alameh [1], Yahya Abbass [1], Ali Ibrahim [1,2,*] and Maurizio Valle [1]

1 Department of Electrical, Electronic and Telecommunication Engineering and Naval Architecture (DITEN)-University of Genoa, via Opera Pia 11a, 16145 Genova, Italy; Mohamad.Alameh@edu.unige.it (M.A.); Yahya.Abbass@edu.unige.it (Y.A.); Maurizio.Valle@unige.it (M.V.)

2 Department of Electrical and Electronics Engineering, Lebanese International University (LIU), Beirut 1105, Lebanon

* Correspondence: Ali.Ibrahim@edu.unige.it; Tel.:+39-3279364917

Received: 1 December 2019; Accepted: 15 January 2020; Published: 18 January 2020

Abstract: Embedding machine learning methods into the data decoding units may enable the extraction of complex information making the tactile sensing systems intelligent. This paper presents and compares the implementations of a convolutional neural network model for tactile data decoding on various hardware platforms. Experimental results show comparable classification accuracy of 90.88% for Model 3, overcoming similar state-of-the-art solutions in terms of time inference. The proposed implementation achieves a time inference of 1.2 ms while consuming around 900 μJ. Such an embedded implementation of intelligent tactile data decoding algorithms enables tactile sensing systems in different application domains such as robotics and prosthetic devices.

Keywords: tactile sensing systems; embedding intelligence; convolutional neural network

1. Introduction

Embedding intelligence near the sensor location may enable tactile sensing systems to be incorporated in many application domains such as prosthetics, robotics, and the Internet of Things. Tactile sensing systems are composed of three main parts, as shown in Figure 1. The distributed tactile sensors are in charge of converting the mechanical stimuli applied on their surface into electrical signals. Tactile sensors could be made from different materials, e.g., capacitive, piezoelectric, and piezoresistive materials [1]; they should be able to enable capabilities similar to what happens on the human skin such as normal and shear force detection, vibration detection, softness, texture, shapes, etc. The readout electronics interface with the sensor arrays by acquiring and digitizing the electrical signals to be then processed by the digital tactile data processing unit [2].

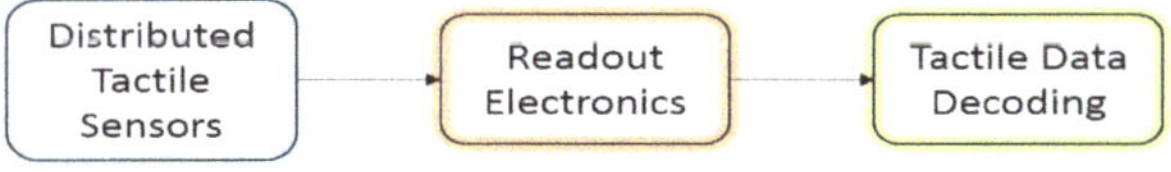

Figure 1. Block diagram of the tactile sensing system.

Decoding tactile information concerns different kinds of tasks, which could be categorized as: simple or complex processing depending on the algorithm's complexity. For simple processing, an example of the information retrieved is temperature, the intensity of the contact force, and contact location, direction, and distribution. Concerning complex processing, more intelligent tasks are expected such as patterns, textures, and roughness classification or touch modalities' discrimination. Employing the complex processing approach enables intelligence in tactile sensing systems. It is achieved by applying sophisticated and complex data decoding algorithms able to extract the meaningful information from sensors. Machine learning (ML) has emerged as an efficient method in many fields and in everyday tasks in smartphones and electronic systems. ML is a powerful learning from examples paradigm used to address classification and regression problems. In particular, Convolutional (CNN) and Deep Neural Networks (DNN) have recently proven their effectiveness when applied to image recognition and tactile data decoding [3]. Many recent research works have focused on the development of ML algorithms for tactile sensing systems [4]. However, embedding machine learning algorithms on hardware platforms near the sensors location is challenging due to the complexity such algorithms impose in terms of time latency and energy consumption. Our main goal is to achieve a tactile sensing system able to perform smart tasks. This system is intended to be portable/wearable, for which the energy budget is limited. Moreover, for the target applications, i.e., robotics and prosthetics, being lightweight is a critical constraint limiting the hardware and battery size.

In this perspective, this paper presents the implementation of CNN algorithms on different hardware platforms. The main contribution of this paper may be summarized as follows:

- It proposes an optimized CNN model, adopted from Gandarias et al.'s research [5], based on reduced data, which demonstrates the ability to provide comparable results in terms of accuracy, i.e., 90.88%, with reduced hardware complexity.
- It presents efficient implementations of the CNN model on different hardware platforms for embedded tactile data processing. The proposed implementations achieve a time inference of 1.2 ms while consuming around 900 μJ. The work demonstrates its suitability for real-time embedded tactile sensing systems.
- It raises a discussion about integrating intelligence into tactile sensing systems and how it enables tactile sensing systems in different application domains.

The remainder of this paper is organized as follows: Section 2 reports the state-of-the-art, showing the recent embedded CNN implementations. In Section 3, we illustrate the experimental setup and methodology. In Section 4, the hardware implementation is explained. The results and discussion are presented in Section 5, followed by the conclusions in Section 6.

2. State-of-the-Art

Different works have addressed the tactile data classification problem, using different methods including, but not limited to, machine learning and deep learning [6–10]. While most of the work done was focused on the methodology itself, few works addressed the implementation on embedded platforms where the real application should reside. Gandarias et al. [11] used two approaches to classify eight objects: finger, hand, arm, pen, scissors, pliers, sticky tape, and Allen key, using a 28×50 tactile sensory array attached to a robotic arm, the first approach using the Speeded-Up Robust Features (SURF) descriptor, while the second a pre-trained AlexNet CNN for feature extraction, with a Support Vector Machine (SVM) classifier for both approaches. In Yuan et al.'s research [12], a CNN was also used for active tactile clothing perception, to classify clothes grasped by a robotic arm equipped with a tactile sensor that output a large RGB pressure map. Based on different textile properties: thickness, smoothness, textile type, washing method, softness, stretchiness, durability, woolen, and wind proof. Each property held two or more labels, e.g., the thickness can be a number from 0–4, and the employed model for textile classification was VGG-19 pretrained on ImageNet [13]. In Rouhafzay et al. [14], a combination of virtual tactile sensors and visual guidance was employed to distinguish eight classes of simulated objects; the tactile sensor size was 32×32, and the input size of the neural network was $32 \times 32 \times 32$, which was a sequence of tactile sensor images. Abderrahmane et al. [15] introduced a zero-shot object recognition framework, to identify previously unknown objects based on haptic feedback. They used BioTac sensors, and two CNNs were employed: one for visual data (input size: $224 \times 224 \times 30$) and the other for tactile data (32×30). They overcame the results of SVM in a previous work [16]. In Alameh et al.'s research [3], transfer learning was used to classify touch modalities obtained through a small 4 × 4 piezoresistive sensory array, by transforming tensorial data into images and then using different CNN models trained on ImageNet [13]. In Gandarias et al.'s research [5], they used a light CNN based (only three convolutional layers inside) on AlexNet, to identify 22 objects using their pressure map, collected from a 28 × 50 tactile sensory array. Other works include those in [17–19].

While all these previous works were not implemented in an embedded environment, we can find few others targeting an embedded implementation for tactile sensing applications. The need for embedded implementation arises from the need to have low power, small form factor electronics to process the tactile information, especially in prosthetic applications [20]. Osta et al. [21] demonstrated an energy efficient system for binary touch modality classification, based on SVM and implemented on a custom hardware architecture. The energy per inference was 81 mJ, and the inference time was 3.3 s. Ibrahim et al. [22] presented a real-time implementation on FPGA for touch modality classification. Using SVM, they achieved a 350 ms inference time and a 945 mJ inference energy for three class classification, as well as 970 ms/6.01 J for a five class classification.

3. Experimental Setup and Methodology

3.1. Dataset

Targeting the classification of tactile data, the use of the dataset collected in [5] was considered. Tactile data were collected by a high resolution (1400 pressure taxels) tactile array, which was attached to the 6 DOF robotic arm AUBO Our-i5 [5]. A set of piezoresistive tactile sensors was distributed with a density of 27.6 taxels/cm^2, forming a matrix of 28 rows by 50 columns. The dataset was composed of pressure images that presented the compliance of 22 objects with the tactile sensors. These images were divided into 22 classes labeled as adhesive, Allen key, arm, ball, bottle, box, branch, cable, cable pipe, caliper, can, finger, hand, highlighter pen, key, pen, pliers, rock, rubber, scissors, sticky tape, and tube. Figure 2 shows an example of the tactile images of three objects used for the training of the CNN model. Each taxel in the tactile array presents a pixel in the pressure image; thus, each pressure image is 28 × 50 × 3 in size. Therefore, the color of the pixel presents the pressure applied at the corresponding taxel. The minimum pressure is presented by black color, and the maximum pressure is presented by red color. Pressure images were then transformed into grayscale images (image size = 28 × 50 × 1), forming the tactile dataset.

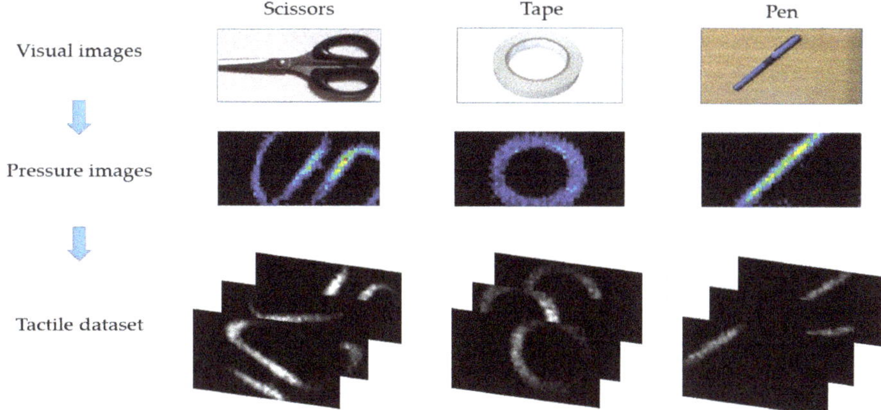

Figure 2. Examples of visual (**top**) vs. pressure (**middle**) vs. tactile images (**bottom**) of common objects.

3.2. Tested Model

Due to computational and memory limitations in the embedded application, a light CNN model was required to perform classification tasks with high accuracy and fewer parameters. In this work, we chose to use one of the models implemented in [5] as a base model to classify the objects in the aforementioned dataset. Among all the implemented networks, we chose to use the custom network TacNet4 because it was the best network that fit the embedded application (fewer parameters with high accuracy [5]). The model was based on AlexNet, which is usually used in computer vision for object classification [23]. The network was composed of 3 Convolutional layers (Conv1, Conv2, and Conv3) with filters sizes (5 × 5, 8), (3 × 3, 16), and (3 × 3, 32) respectively. Each convolutional layer was followed by a Batch Normalization (BaN), Activation (ReLU), and Maxpooling (Maxpool) layer, respectively, where all pooling layers used 2 × 2 maxpooling with a stride of two. A Fully Connected layer (FC = [fc4]) with 22 neurons followed by a softmax layer were used to classify the input tactile data and give the likelihood of belonging to each class (object). The input shape of the model was configured to the size of the collected tactile data. Figure 3 shows the detailed structure of the network used.

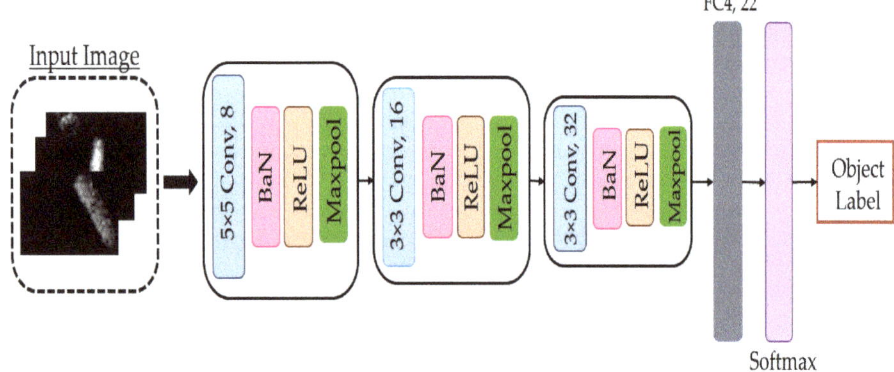

Figure 3. Architecture of the tested model. BaN, Batch Normalization.

The network was implemented in MATLAB R2019b using the Neural Network Toolbox. A total of 1100 tactile images were used to train the model. The learning process was implemented in MATLAB by dividing the tactile data into three sets: training, validation, and test sets.

When having an adequate dataset, the validation set is expected to be a good statistical representation of the entire dataset. If not, the results of the training procedure highly depend on how the dataset is divided.

To avoid this, In this work, we used the cross-validation method. The data were partitioned into five folds, and each fold was divided into training, validation, and test sets. The training set formed 80% of the dataset, and the validation and test sets formed 10% each. This process was then repeated five times until all the folds were used, without having common elements across all folds for the validation and test sets, as shown in Figure 4.

For each training process, the training set was composed of 880 images, 40 images for each label, whilst each of the validation and test sets was composed of 110 images. Training the model from scratch required a large dataset to achieve high accuracy. For that reason, data augmentation techniques, i.e., flipping, rotation, and translation in the X and Y axis, were applied to the dataset. Hence, the amount of tactile data available for training and validation was increased to 5280 and 660, respectively. The performance of the implemented model was evaluated based on the recognition rates achieved in a classification experiment of the test set composed of 110 original images (objects) from 22 classes.

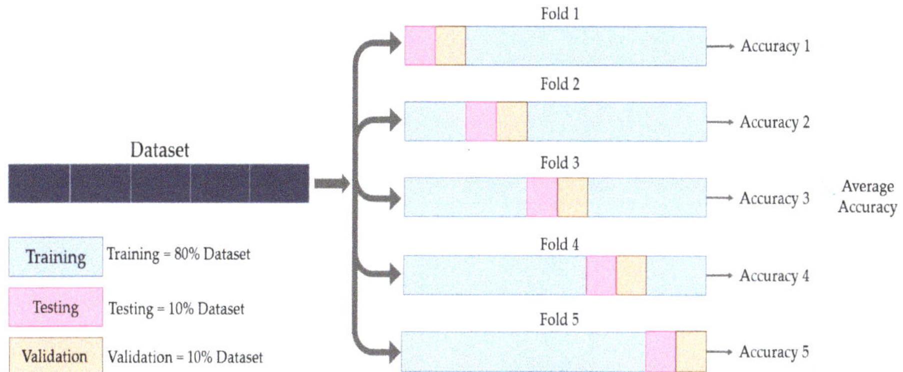

Figure 4. Visual representation of the training, test, and validation split using cross-validation.

For embedded applications, with computational, memory, and energy constraints, it is necessary to decrease the number of trainable parameters in the CNN model. In this work, we chose to decrease the number of parameters of the trained model by decreasing the input image size (i.e., lower resolution images); an example is shown in Figure 5. For that reason, several experiments were performed to choose the smaller size of the input data, keeping the same classification accuracy. The input shapes were chosen in a way that each shape resulted in a reduction of the number of parameters.

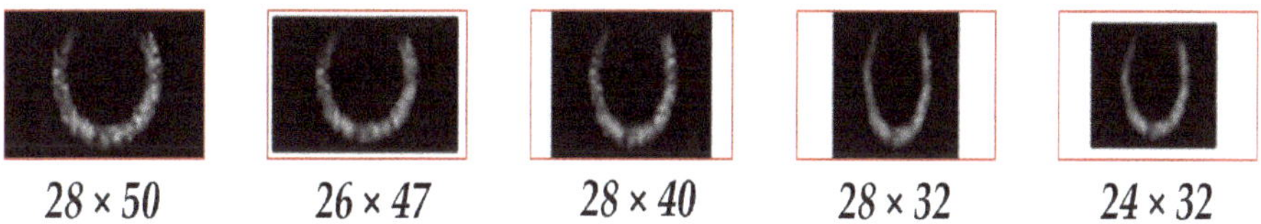

Figure 5. Example of an image resized for the sticky tape object; the red canvas is shown for illustration, which signifies the original image size (28 × 50).

Table 1 shows how the number of parameters of the layers depended on the input shape. The change in the input shape affected only the number of parameters of the fully connected layer. This was due to the fact that the number of parameters in the convolutional layer depended only on the size and number of the filters assigned for each layer (((width of the filter × height of the filter) + 1) × No. of filters), while in the FC layer, the number of parameters ((No. of neurons in the FC layer × No. of neurons in the previous layer) + 1) was affected by the size of the input image and the output layer. The performance of the model was studied with five different input shapes, as shown in Table 1. This resulted in five different models with different input shapes, each one trained from scratch 5 times (one time per fold), which output 25 trained NNs. Figure 6 shows the training and validation accuracy over epochs, for the first three models among the five models. The figure shows that the accuracy achieved by the three models was close to 100%. Each model was evaluated with MATLAB by running a classification task on the test set.

Table 1. Distribution of the number of parameters on the models' layers.

Layers	Model 1 (28 × 50)	Model 2 (26 × 47)	Model 3 (28 × 40)	Model 4 (28 × 32)	Model 5 (24 × 32)
Conv1	208	208	208	208	208
BaN1	16	16	16	16	16
Conv2	1168	1168	1168	1168	1168
BaN2	32	32	32	32	32
Conv3	4640	4640	4640	4640	4640
BaN3	64	64	64	64	64
FC	19,734	16,918	14,102	11,286	8470
Total	25,862	23,046	20,230	17,414	14,598

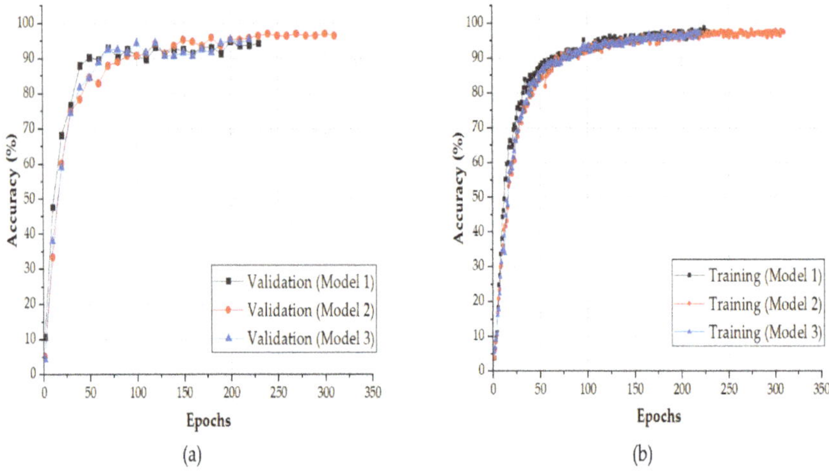

Figure 6. Learning accuracy for the 3 configurations of the TactNet4model: (**a**) training; (**b**) validation.

Figure 7 shows the change in the number of trainable parameters and the average classification accuracy, with respect to the change in the input shape, as well as the FLOPs. The classification accuracy presented the average test accuracy among the five folds. The figure shows that it was possible to decrease the input size from 28 × 50 × 1 to 26 × 47 × 1 or to 28 × 40 × 1 and achieve an increase in the classification accuracy from 90.70% to 91.98% and 90.88%, respectively. Decreasing the input size of the model resulted in a drop in the trainable parameters from 25,862 to 23,046 and 20,230 parameters, respectively, for the aforementioned models. This decrease in the number of parameters would also induce a decrease of the number of Floating Point Operations (FLOPs), as shown in Figure 7; the average ratio of the decrease in the number of parameters with respect to the decrease in the number of FLOPs was 1/44 i.e., with each decrease in number of parameters, there was a 44 times decrease of the FLOPs. The number of FLOPS in Figure 7 corresponds to the convolutional layers only, where most of the FLOPs were, and these FLOPs were calculated according to the following formula [24]: FLOPs = n × m × k, where n is the number of kernels, k is the size of the kernel (width × height × depth), and m the size of output feature map (width × height), while the depth in the kernel size corresponds to the depth of the input feature map.

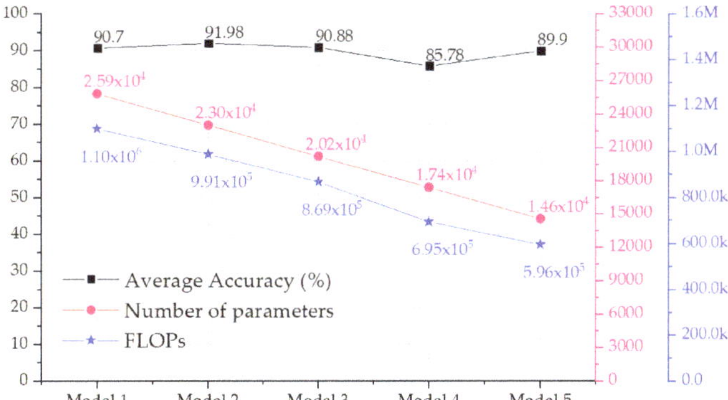

Figure 7. Comparison of the performance, number of trainable parameters, and FLOPS in the convolutional layers.

4. Embedded Hardware Implementations

The models obtained from MATLAB were converted to Open Neural Network Exchange (ONNX) format [25]. ONNX provides an open source format for AI models, both deep learning and traditional ML, which enables the inter-operability between different frameworks. Figure 8 shows how the CNN model was converted into different formats for different hardware platforms. Figure 7 shows the number of trainable parameters and the corresponding accuracy for each model. It is clearly shown that all models preserved comparable accuracy, but the best were the first three, i.e., Model 1, Model 2, and Model 3. However, since Model 2 and Model 3 demonstrated a reduced number of training parameters and accordingly a reduced number of operations (FLOPS), they were selected for the hardware implementation. This choice was based on the fact that reducing FLOPS reduced the inference time and power consumption.

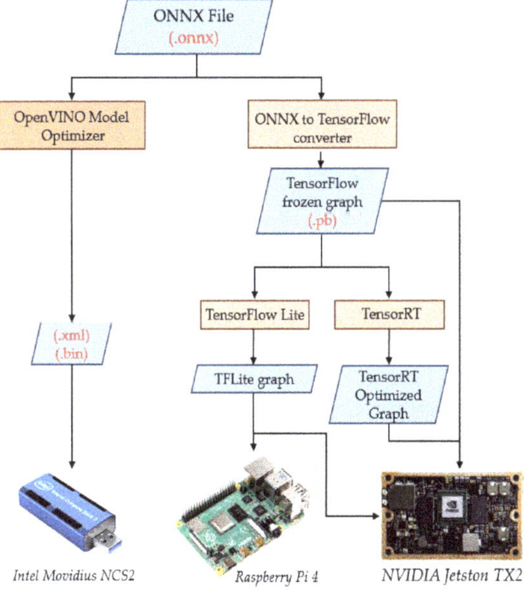

Figure 8. Implementation flow.

The reason behind the selection of hardware platforms was as follows:

1. The custom architecture targeting the embedded implementation of neural networks, e.g., Movidius NCS2.
2. The high usability of ARM processors in embedded architectures, e.g., Raspberry Pi 4.
3. The high performance architecture, designed for parallel processing in general, and also optimized for embedded applications: e.g., NVidia Jetson TX2.
4. The support for the execution of pretrained neural network models coming from different platforms without retraining.

4.1. Movidius Neural Compute Stick 2

Movidius Neural Compute Stick 2 (NCS2) is a hardware accelerator designed by Intel for on-chip neural network inference, especially CNNs, equipped with the Intel Movidius MyriadX Vision Processing Unit (VPU). It contains 16 SHAVE (Streaming Hybrid Architecture Vector Engine) cores [26] and a dedicated hardware neural network accelerator. It requires a host to flash the neural network, as well as to feed it with data and invoke the inference to get the results back via the USB 3.0 port. The host can be a Linux, Windows, or Mac based machine. To achieve these tasks, Intel provides OpenVINO: Open Visual Inference and Neural network Optimization Toolkit, a cross platform toolkit that enables deep learning inference and easy heterogeneous execution across multiple Intel® hardware (VPU, GPU, CPU, FPGA). The optimizations offered by OpenVINO are: batch normalization and scale shift, linear operation merge and linear operation fusion. The details were mentioned in [27].

4.2. Jetson TX2

NVidia's Jetson TX2 [28] is a power efficient embedded AI computing device, designed mainly for edge AI, and belongs to the Pascal™ family of GPUs, loaded with 8 GB of memory, 59.7 GB/s of memory bandwidth, and 8 GB of RAM. In this experiment, we used TensorFlow (TF) [29] for the inference, as well as NVidia TensorRT [30] under Ubuntu OS. TF is an open source end-to-end machine learning platform, while TensorRT is a platform for high performance deep learning inference dedicated to NVidia hardware. It includes a deep learning inference optimizer and a runtime that delivers low latency and high throughput for deep learning inference applications.

As an optimization for TensorFlow, TensorFlow Lite (TFLite) [31] is an open source deep learning framework for on device inference. The same TensorFlow model can be converted into the TFLite model. To perform an inference with a TFLite model, the TFLite interpreter is required, which uses a static graph ordering and a custom (less dynamic) memory allocator to ensure minimal load, initialization, and execution latency [31], also reducing the weights' precision, e.g., floating point vs. fixed point precision, without affecting the accuracy.

4.3. ARM

As for the implementation on the ARM architecture, we used Raspberry Pi 4, equipped with a Quad core Cortex-A72 (ARM v8) 64-bit System on Chip (SoC) @ 1.5 GHz and 4 GB RAM. For the inference on this hardware, we used the TFLite runtime library, under the Ubuntu OS.

For all the mentioned platforms, both power consumption and inference time were calculated. The inference time was calculated by averaging 110 inferences, which corresponded to the test set size. As for the power consumption, two methods were used:

1. the provided APIs in Jetson TX2, which provided readings about voltage, power, and input current to the GPU.
2. the external USB multimeter, connected in serial to the power source for both Raspberry Pi and the Movidius NCS2.

5. Results and Discussion

In this work, we achieved a better accuracy in tactile data classification using CNN compared to the original model obtained in [5], even by resizing the input, therefore decreasing the number of trainable parameters. The chosen models reduced the number of trainable parameters by a maximum of 21.77% of the original trainable parameters and also increased the accuracy by a maximum of 1.28%, noting that Model 5 (24 × 32 × 1) with 0.8% less accuracy than the original model had 42% fewer trainable parameters. Choosing the right model depended on the implementation, i.e., a trade-off between accuracy and hardware complexity should take place: if the best accuracy was targeted, then Model 2 should be selected; while the choice of Model 3 would be when less hardware complexity was needed, accepting a small accuracy degradation. Reducing the input size while still keeping the same or even better accuracy could be explained in three points:

1. The random initialization of the weights may lead in different runs to different accuracy results, e.g., 10 different runs for training Fold 4 of Model 2 with the same hyperparameters gave different results, as shown in Table 2, which shows an average of 94.36% and a standard deviation of 1.904%.
2. Random selection of batch data and data shuffling would affect also the update of the weights and make them different from one training to another.
3. The feature extraction process achieved by CNN was error resilient [32]. A CNN could still extract features even with some manipulation of the input image. This was one of the reasons for data augmentation [33] when training neural networks, which was to let the neural network learn the features even from augmented images (scaled, rotated, flipped, etc.), instead of learning only the samples in the original dataset. In our case, the features were still detectable even after image resizing, as shown in Figure 5.

Table 2. Accuracy results for 10 runs on Model 2, Fold 4.

Trials	Accuracy (%)
1	96.36
2	92.73
3	94.55
4	91.82
5	97.27
6	93.64
7	92.73
8	95.45
9	96.36
10	92.73
Average ± Stdev	94.36 ± 1.904%

According to Tables 3 and 4, the smallest power consumption and inference time were obtained using TensorRT under Jetson TX2, which was 153 mW dynamic power within 5.29 ms as the inference time, implying 0.809 × 10 $^{-3}$ Joules of dynamic energy (see Table 5). The most dynamic energy consumption was for the Intel Movidius NCS2, 1.9 ms × 800 mW = 1.52 × 10 $^{-3}$ Joules. Regarding the power consumption, since the neural network used was small compared to the hardware capacity, the power consumption was almost the same for the three models, noting that the accuracy on the USB power meter was on the 10 mW scale, so that a difference of less than 10 mW between two measurements could not be detected using this instrument.

Table 3. Comparison of the inference time between models.

Platform		Inference Time (ms)		
Hardware	**Software**	**Model 1**	**Model 2**	**Model 3**
Jetson TX2	TensorRT	5.5597	5.2905	5.919
	TF	6.2943	5.4691	5.946
	TFLite	1.3384	1.2181	1.2445
Core i7	MATLAB	3.245	2.6139	2.4715
Movidius NCS2	OpenVINO	1.9	1.9	1.86
Raspberry Pi4	TFLite	1.615	1.473	1.21

Table 4. Power consumption.

Platform		Current (mA)		Voltage (V)	Consumed Power (mW)		
Hardware	**Software**	**Static**	**Total**		**Static**	**Total**	**Dynamic**
Jetson	TensorRT	8	16	19.072	152	305	153
	TF	8	16	19.072	152	305	153
Movidius NCS2	OpenVINO	-	160	5	-	800	800
Raspberry Pi4	TFLite	560	700	5	2800	3500	700

Table 5. Energy consumption.

Platform		Energy Consumption (μJ)		
Hardware	**Software**	**Model 1**	**Model 2**	**Model 3**
Jetson TX2	TensorRT	850.6341	809.4465	905.607
	TF	963.0279	836.7723	909.738
Movidius NCS2	Open VINO	1520	1520	1488
Raspberry Pi4	TFLite	1130.5	1031.1	847

6. Conclusions

This paper presented the implementation of a smart tactile sensing system based on an embedded CNN approach. The proposed model optimized a state-of-the-art model proposed in [5] by reducing the input data size. The experimental results were comparable in terms of accuracy after reducing the size from (28 × 50) to (26 × 47) and (28 × 40). The hardware implementation on different hardware platforms, namely Movidius NCS2, NVidia's Jetson TX2, and Cortex-A72 (ARM v8), was provided. The proposed models showed better performance on hardware platforms when time inference was compared. Power consumption was also measured and compared among different platforms. Targeting portable tactile sensing systems, the proposed work demonstrated the feasibility of integrating machine learning methods on a hardware platform to enable intelligence for such a system. This may pave the way for smart tactile sensing systems to be applied in prosthetics and robotics.

Author Contributions: Conceptualization, A.I.; methodology, M.A. and A.I.; software, M.A. and Y.A; validation, A.I.; investigation, M.A. and Y.A.; data curation, M.A. and Y.A.; writing, original draft preparation, Y.A., M.A., and A.I.; writing, review and editing, A.I. and M.V.; visualization, Y.A.; supervision, A.I. and M.V.; funding acquisition, M.V. All authors read and agreed to the published version of the manuscript.

Funding: The authors acknowledge financial support from Compagnia di San Paolo, Grant Number 2017.0559, ID ROL19795.

Acknowledgments: The authors would like to thank the NVidia corporation for the donation of the Jetson TX2 development kit.

Conflicts of Interest: The authors declare no conflict of interest.

Abbreviations

The following abbreviations are used in this manuscript:

CNN	Convolutional Neural Network
DCNN	Deep Convolutional Neural Network
SVM	Support Vector Machine
ML	Machine Learning
FPGA	Field-Pogrammable Gate Array

References

1. Ibrahim, A.; Pinna, L.; Seminara, L.; Valle, M. Achievements and Open Issues Toward Embedding Tactile Sensing and Interpretation into Electronic Skin Systems. In *Material-Integrated Intelligent Systems-Technology and Applications*; John Wiley & Sons, Ltd.: WeinHeim, Germany, 1 December 2017; Chapter 23, pp. 571–594. doi:10.1002/9783527679249.ch23. [CrossRef]
2. Saleh, M.; Abbass, Y.; Ibrahim, A.; Valle, M. Experimental assessment of the interface electronic system for PVDF-based piezoelectric tactile sensors. *Sensors* **2019**, *19*, 4437. doi:10.3390/s19204437. [CrossRef] [PubMed]
3. Alameh, M.; Ibrahim, A.; Valle, M.; Moser, G. DCNN for Tactile Sensory Data Classification based on Transfer Learning. In Proceedings of the 2019 15th Conference on Ph.D Research in Microelectronics and Electronics (PRIME), Lausanne, Switzerland, 15–18 July 2019; pp. 237–240. doi:10.1109/prime.2019.8787748. [CrossRef]
4. Luo, S.; Bimbo, J.; Dahiya, R.; Liu, H. Robotic tactile perception of object properties: A review. *Mechatronics* **2017**, *48*, 54–67. doi:10.1016/j.mechatronics.2017.11.002. [CrossRef]
5. Gandarias, J.M.; Garcia-Cerezo, A.J.; Gomez-de Gabriel, J.M. CNN-Based Methods for Object Recognition With High-Resolution Tactile Sensors. *IEEE Sens. J.* **2019**, *19*, 6872–6882. doi:10.1109/jsen.2019.2912968. [CrossRef]
6. Cheng, G.; Dean-Leon, E.; Bergner, F.; Olvera, J.R.G.; Leboutet, Q.; Mittendorfer, P. A Comprehensive Realization of Robot Skin: Sensors, Sensing, Control, and Applications. *Proc. IEEE* **2019**, *107*, 2034–2051. doi:10.1109/JPROC.2019.2933348. [CrossRef]
7. Martinez-Hernandez, U.; Dodd, T.J.; Prescott, T.J. Feeling the Shape: Active Exploration Behaviors for Object Recognition With a Robotic Hand. *IEEE Trans. Syst. Man Cybern. Syst.* **2018**, *48*, 2339–2348. doi:10.1109/TSMC.2017.2732952. [CrossRef]
8. Zou, L.; Ge, C.; Wang, Z.; Cretu, E.; Li, X. Novel tactile sensor technology and smart tactile sensing systems: A review. *Sensors* **2017**, *17*, 2653. [CrossRef] [PubMed]
9. Li, R.; Adelson, E.H. Sensing and Recognizing Surface Textures Using a GelSight Sensor. In Proceedings of the 2013 IEEE Conference on Computer Vision and Pattern Recognition, Portland, OR, USA, 23–28 June 2013; IEEE: Portland, OR, USA, 2013; pp. 1241–1247. doi:10.1109/CVPR.2013.164. [CrossRef]
10. Schmitz, A.; Bansho, Y.; Noda, K.; Iwata, H.; Ogata, T.; Sugano, S. Tactile object recognition using deep learning and dropout. In Proceedings of the 2014 IEEE-RAS International Conference on Humanoid Robots, Madrid, Spain, 18–20 November 2014; pp. 1044–1050. doi:10.1109/HUMANOIDS.2014.7041493. [CrossRef]
11. Gandarias, J.M.; Gomez-de Gabriel, J.M.; Garcia-Cerezo, A. Human and object recognition with a high-resolution tactile sensor. In Proceedings of the 2017 IEEE SENSORS, Glasgow, UK, 29 October–1 November 2017; IEEE: Glasgow, 2017; pp. 1–3. [CrossRef]
12. Yuan, W.; Mo, Y.; Wang, S.; Adelson, E. Active Clothing Material Perception using Tactile Sensing and Deep Learning. *arXiv* **2017**, arXiv:1711.00574.
13. ImageNet. Available online: http://www.image-net.org (accessed on 20 November 2019).
14. Rouhafzay, G.; Cretu, A.M. An Application of Deep Learning to Tactile Data for Object Recognition under Visual Guidance. *Sensors* **2019**, *19*, 1534. doi:10.3390/s19071534. [CrossRef] [PubMed]
15. Abderrahmane, Z.; Ganesh, G.; Crosnier, A.; Cherubini, A. Visuo-Tactile Recognition of Daily-Life Objects Never Seen or Touched Before. In Proceedings of the 2018 IEEE 15th International Conference on Control, Automation, Robotics and Vision (ICARCV), Singapore, 18–21 November 2018; pp. 1765–1770.

16. Abderrahmane, Z.; Ganesh, G.; Crosnier, A.; Cherubini, A. Haptic Zero-Shot Learning: Recognition of objects never touched before. *Robot. Auton. Syst.* **2018**, *105*, 11–25. doi:10.1016/j.robot.2018.03.002. [CrossRef]
17. Li, J.; Dong, S.; Adelson, E. Slip detection with combined tactile and visual information. In Proceedings of the 2018 IEEE International Conference on Robotics and Automation (ICRA), Brisbane, QLD, Australia, 21–25 May 2018; pp. 7772–7777.
18. Wu, H.; Jiang, D.; Gao, H. Tactile motion recognition with convolutional neural networks. In Proceedings of the 2017 IEEE/RSJ International Conference on Intelligent Robots and Systems (IROS), Vancouver, BC, Canada, 24–28 September 2017; pp. 1572–1577.
19. Kwiatkowski, J.; Cockburn, D.; Duchaine, V. Grasp stability assessment through the fusion of proprioception and tactile signals using convolutional neural networks. In Proceedings of the 2017 IEEE/RSJ International Conference on Intelligent Robots and Systems (IROS), Vancouver, BC, Canada, 24–28 September 2017; IEEE: Vancouver, BC, Canada, 2017; pp. 286–292. doi:10.1109/IROS.2017.8202170. [CrossRef]
20. Fares, H.; Seminara, L.; Ibrahim, A.; Franceschi, M.; Pinna, L.; Valle, M.; Dosen, S.; Farina, D. Distributed Sensing and Stimulation Systems for Sense of Touch Restoration in Prosthetics. In Proceedings of the 2017 New Generation of CAS (NGCAS), Genova, Italy, 6–9 September 2017; pp. 177–180. [CrossRef]
21. Osta, M.; Ibrahim, A.; Magno, M.; Eggimann, M.; Pullini, A.; Gastaldo, P.; Valle, M. An Energy Efficient System for Touch Modality Classification in Electronic Skin Applications. In Proceedings of the 2019 IEEE International Symposium on Circuits and Systems (ISCAS), Sapporo, Japan, 26–29 May 2019; pp. 1–4.
22. Ibrahim, A.; Gastaldo, P.; Chible, H.; Valle, M. Real-time digital signal processing based on FPGAs for electronic skin implementation. *Sensors* **2017**, *17*, 558. [CrossRef] [PubMed]
23. Jansen, K.; Zhang, H. Scheduling malleable tasks.May 23, 2018 In *Handbook of Approximation Algorithms and Metaheuristics*; Chapman and Hall/CRC: New York, NY, USA, 15 May 2007; pp. 45-1–45-16. doi:10.1201/9781420010749. [CrossRef]
24. Lu, Z.; Rallapalli, S.; Chan, K.; La Porta, T. Modeling the resource requirements of convolutional neural networks on mobile devices. In Proceedings of the MM 2017—Proceedings of the 2017 ACM Multimedia Conference, Mountain View, CA, USA, 23–27 October 2017; pp. 1663–1671, doi:10.1145/3123266.3123389. [CrossRef]
25. Open Neural Network Exchange. Available online: https://github.com/onnx/onnx/ (accessed on 20 November 2019).
26. Intel Movidius NCS2. Available online: https://software.intel.com/en-us/neural-compute-stick (accessed on 20 November 2019).
27. OpenVino Model Optimization Techniques. Available online: https://docs.openvinotoolkit.org/latest/_docs_MO_DG_prepare_model_Model_Optimization_Techniques.html (accessed on 20 November 2019).
28. NVIDIA Jetson Modules and Developer Kits for Embedde Systems Development. Available online: https://www.nvidia.com/en-us/autonomous-machines/embedded-systems (accessed on 20 November 2019).
29. TensorFlow. Available online: https://www.tensorflow.org (accessed on 20 November 2019).
30. NVIDIA TensorRT. Available online: https://developer.nvidia.com/tensorrt (accessed on 20 November 2019).
31. TensorFlow Lite. Available online: https://www.tensorflow.org/lite (accessed on 20 November 2019).
32. Hanif, M.A.; Hafiz, R.; Shafique, M. Error resilience analysis for systematically employing approximate computing in convolutional neural networks. In Proceedings of the 2018 Design, Automation Test in Europe Conference Exhibition (DATE), Dresden, Germany, 19–23 March 2018; pp. 913–916. [CrossRef]
33. Perez, L.; Wang, J. The Effectiveness of Data Augmentation in Image Classification using Deep Learning. *arXiv* **2017**, arXiv: 1712.04621.

Article

Implementation of Hand Gesture Recognition Device Applicable to Smart Watch Based on Flexible Epidermal Tactile Sensor Array

Sung-Woo Byun [1] and Seok-Pil Lee [2,*]

1 Department of Computer Science, Graduate School, SangMyung University, 20, Hongjimun 2-gil, Jongno-gu, Seoul 03016, Korea; 123234566@naver.com

2 Department of Electronic Engineering, SangMyung University, 20, Hongjimun 2-gil, Jongno-gu, Seoul 03016, Korea

* Correspondence: esprit@smu.ac.kr

Received: 17 August 2019; Accepted: 10 October 2019; Published: 12 October 2019

Abstract: Ever since the development of digital devices, the recognition of human gestures has played an important role in many Human-Computer interface applications. Various wearable devices have been developed, and inertial sensors, magnetic sensors, gyro sensors, electromyography, force-sensitive resistors, and other types of sensors have been used to identify gestures. However, there are different drawbacks for each sensor, which affect the detection of gestures. In this paper, we present a new gesture recognition method using a Flexible Epidermal Tactile Sensor based on strain gauges to sense deformation. Such deformations are transduced to electric signals. By measuring the electric signals, the sensor can estimate the degree of deformation, including compression, tension, and twist, caused by movements of the wrist. The proposed sensor array was demonstrated to be capable of analyzing the eight motions of the wrist, and showed robustness, stability, and repeatability throughout a range of experiments aimed at testing the sensor array. We compared the performance of the prototype device with those of previous studies, under the same experimental conditions. The result shows our recognition method significantly outperformed existing methods.

Keywords: gesture recognition; flexible epidermal tactile sensor array; wearable device; wearable sensors

1. Introduction

Ever since the development of digital devices, the recognition of human gestures has played an important role in many Human-Computer interface (HCI) applications, permitting interaction in a natural and comfortable way [1–4]. Hand gesture recognition has the advantage of being applicable to a range of applications, such as handling presentations, controlling drones, and more [5]. A universal remote-control system using hand gestures is presented in [6]. Hand gesture recognition is achieved using two main kinds of sensors: contact sensors and non-contact sensors. The non-contact methods are primarily based on visual technologies such as camera sensors, Kinect, and Leap Motion controller (LMC), which do not require attaching the sensors to the human body, as reported by various studies [7–13]. Contact methods identify gestures by analyzing the signal acquired from contact sensors, which are wrapped around the user's arm or wrist, or are attached to a glove that the user wears [14–16]. They have a wider recognition range than the non-contact methods, without constraints such as limited range and the sight of sensors, and relatively accurate information can be acquired due to the direct contact with the user. For this reason, various wearable devices have been developed, and inertial sensor, magnetic sensor, gyro sensor, electromyography (EMG), force-sensitive resistors (FSRs), and others have been used to identify gestures. In particular, the EMG sensor has been used in many studies on gesture recognition [17–19]. Many researchers used EMG sensors for recognizing the intention of an operator [17,20–22]. Recently, to control digital devices, Thalmic Labs Co. developed an EMG-based gesture recognition device, which is referred to as Myo [5]. The device was designed as an armband bracelet to measure EMG signals from the forearm muscles. EMG-based methods have become more important in the practical application of surface electromyography [22]. The main challenges of EMG-based methods are the weak signal intensity with noise [23]. Generally, the amplitude range is 0–105 mV and the bandwidth is 0.5–2 kHz, so it is easily interfered in by the external noise and the acquisition device itself [22,24].

Another gesture recognition approach is the use of FSRs. FSRs sensors detect muscle activity by measuring and monitoring changes in resistance generated by movements of the muscles [25,26]. Since the muscle contraction occurs the longitudinal elongation and the expansion of its cross-sectional area, it is possible to detect the muscular activity by monitoring the swelling of muscles by FSRs sensor [24]. FSRs sensor is robust to noise compared to the other bio signal measurements, but the output voltage of FSRs sensors is nonlinear due to relationship between output voltage and resistance [25]. In addition, since FSRs sensor is a thin film, and thus an input device with FSRs sensor should become rigid, which causes discomfort in wearing [24].

Mechanomyography (MMG) can also be used to detect muscular activities. Muscular activity is identified by mechanical vibration, which is generated by the tremor of each muscle fiber [24]. MMG-based methods commonly use an accelerometer [27,28] and a microphone [29,30]. However, MMG based on an accelerometer can only be used when the magnitude of acceleration is distinguishable compared to acceleration due to gravity and motion. MMG based on a sound transducer is reliable only in a silent space [24].

These sensors have been shown to be successful in many studies over the past two decades. However, there are still different drawbacks for each sensor, which affect the detection of gestures. To overcome these problems and accurately recognize gestures, we developed a novel gesture recognition method using a Flexible Epidermal Tactile Sensor Array (FETSA) based on strain gauges to sense deformations. Such deformations are transduced to electric signals. By measuring the electric signals, the sensor array can estimate the degree of deformations, including compression, tension, and twist caused by movements of the wrist. The principle of FETSA is similar to that of MMG sensors and FSR sensors, but its flexibility provides enhanced usability in terms of wearing the sensor. The sensor guarantees linearity, in contrast with FSRs sensors. To test the performance of the sensor, we fabricated a prototype clip-type device, and conducted comparison tests using the porotype device. We compared the sensor with a commercial EMG sensor and an FSRs sensor, which are commonly used in gesture recognition studies. Furthermore, we compared the porotype device with previous studies, under the same experimental conditions. We conducted additional experiments using gestures defined in this research. The resulting recognition method significantly outperformed existing methods.

2. Principle of Flexible Epidermal Tactile Sensor Array

When a gesture occurs, the length and thickness of the muscles around the wrist change during concentric contraction and eccentric contraction, changes which are classified as dynamic contraction. For concentric contraction, related muscles shorten and thicken while muscular force is generated. In eccentric contraction the muscles involved lengthen and become thinner. Isometric contraction corresponds to static contraction; there is no change in the muscle length, although the muscles generate force. Isometric contraction occurs when maintaining a posture or holding an object. Therefore, in order to detect hand movements when gestures occur, we must measure the changes in muscles, whether concentric contraction or eccentric contraction. However, because the EMG sensor measures all three types of contractions, devices based on the EMG sensor require additional signal processing to distinguish isometric contraction from the other two contractions of interest. To unambiguously detect concentric contraction and eccentric contraction, we developed a Flexible Epidermal Tactile Sensor Array to measure the movement of muscles in a reliable and convenient way. The proposed sensor is shown in Figure 1a.

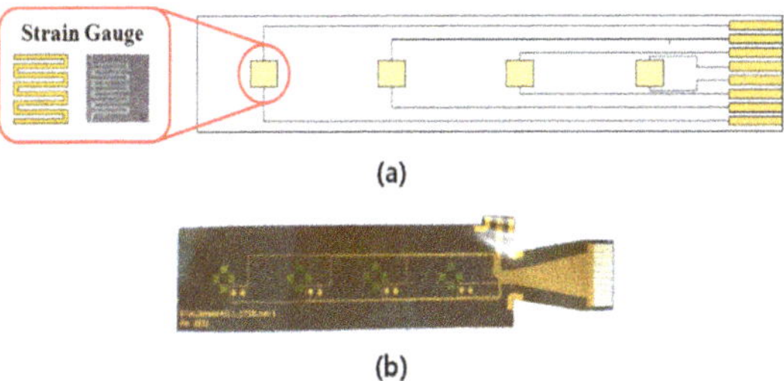

Figure 1. Flexible Epidermal Tactile Sensor Array (FETSA). (**a**) Design of the FETSA. (**b**) Fabricated sensor.

In sensor design, the number and location of sensors are important in order to recognize gestures. The initial model was fabricated with 16-channel sensors, so that it could wrap around the whole of a wrist [31]. Based on preliminary experiments, the final model has four sensors. To detect the movement of wrist muscles, sensors are positioned over the muscles responsible for wrist movements. The device was designed using flexible polyimide, so that it could be worn on the wrist to improve its fit to the user's body surface. Strain gauges are located on the flexible substrate. Figure 1b shows the fabricated sensor array. Depending on the movement of the wrist, the analog resistance value of the flexible array sensor is processed using a circuit and converted into a digital value. This value is then used for gesture recognition.

Figure 2b shows the gesture recognition device based on FETSA. It was designed as a clip so that it could be worn with a smart watch. The data acquisition board includes a serial communication unit, through which the sensor signal is recorded. The baud rate is set to 115200 for real-time processing. Sixty data units per second are acquired through the device. The four sensors of the device detect the activities of muscles responsible for the movement of the wrist as shown in Figure 2a. Channel 1 is located on the abductor pollicis longus muscle, which deals with the up and down movement of the thumb. A sensor is located on each of the muscles responsible for the movement of the wrist.

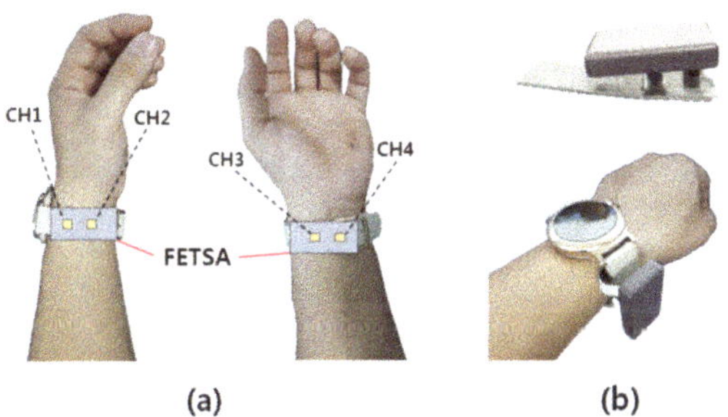

Figure 2. (**a**) Location of sensors. (**b**) How to wear the device.

As discussed above, concentric and eccentric contraction in muscles under the wrist occur when people make hand gestures. When the fist is twisted down, as shown in Figure 3, eccentric contraction occurs at the extensor pollicis brevis muscle. This contraction influences the strain gauge of the sensor located on the muscle. Force generated from the muscle is transmitted to the sensor, raising the resistance value due to the expansion of the sensor. In contrast, when the fist is twisted up, concentric contraction occurs in the muscle, decreasing the resistance value of the sensor. The proposed sensor detects the activities of muscles under the wrist by measuring these deformations of the sensor to detect gestures.

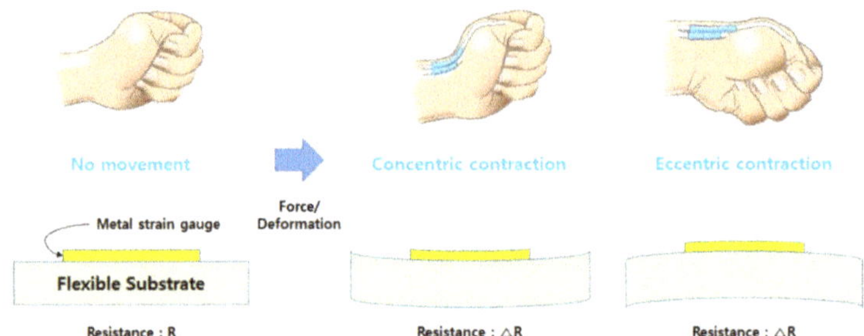

Figure 3. Change of a strain gauge and resistance caused by the movement of a wrist.

3. Gesture Recognition with FETSA

In this section, we explain how the device recognizes gestures. First, we investigated the changes in the signals from each of the sensors according to the movement of the wrist, since the muscles may influence the deformation of adjacent sensors simultaneously. There are eight motions which can be made with wrist and fingers (Figure 4).

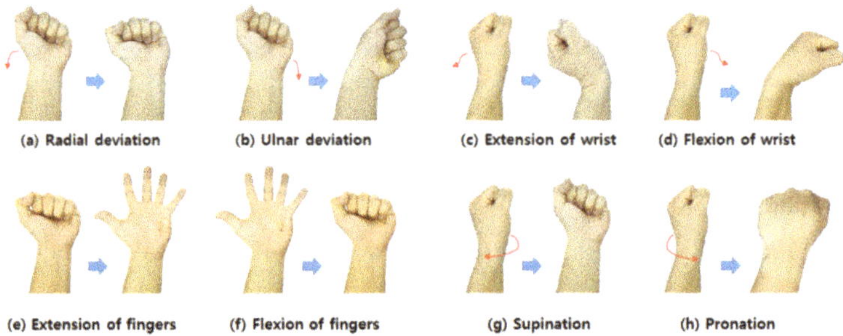

Figure 4. Eight motions of the wrist and fingers: (**a**,**b**) radial and ulnar deviation of the wrist; (**c**,**d**) extension and flexion of the wrist; (**e**,**f**) extension and flexion of fingers; (**g**,**h**) supination and pronation of the wrist.

Since each muscle is theoretically concerned with different motions, each sensor of the device detects different signals according to the motions. We investigated the changes in the signals when a subject made different motions. The subjects started with a light motion by relaxing the hands before making the eight motions shown in Figure 4.

Figure 5 shows the change in each signal when the eight motions shown in Figure 4 occurred. The changes in the signals acquired from the four channels were different in each motion. For instance, in motion (a) and motion (b), the signals acquired from channel 4 and channel 1 are similar, but the signals are different in channel 2 and channel 3 (Figure 5a,b). The proposed method can distinguish the eight motions using the differences in signals, an observation which verifies that the hand gestures can be recognized.

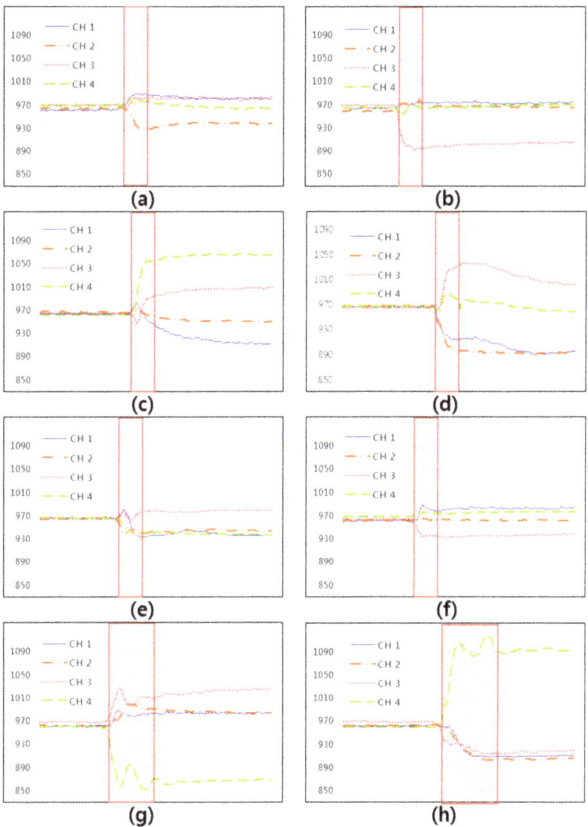

Figure 5. Changes in each of the signals (**a–h**) produced during the eight motions shown in Figure 4. The shaded areas indicate when each movement was made.

The entire process of gesture recognition using the proposed device is shown in Figure 6. The process consists of preprocessing, feature extraction, and classification. The steps are explained in detail as follows.

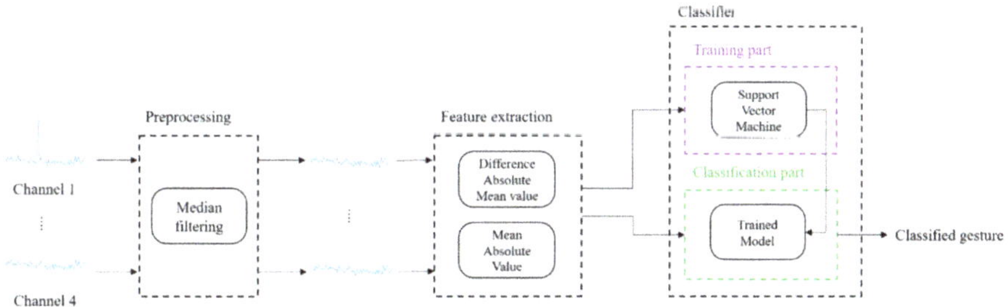

Figure 6. Overview of the process of gesture recognition method using the proposed device.

3.1. Preprocessing

While recording bio-signals, mixed signals sometimes occur due to the presence of noise. For example, noise can be recorded from the heartbeat reflected in the artery under the wrist, and interpreted as a movement of the wrist. Such noise leads to the degradation of the quality of the signal, and must therefore be removed. As apparent in Figure 7a, a recording of a stable signal is periodically deformed by heartbeats. As this deformation can cause reduction in the accuracy of gesture recognition, we used a median filter to remove this noise. This approach is effective in removing impulse noises while preserving the important properties of the signal. Figure 7b shows the results after the noise is removed.

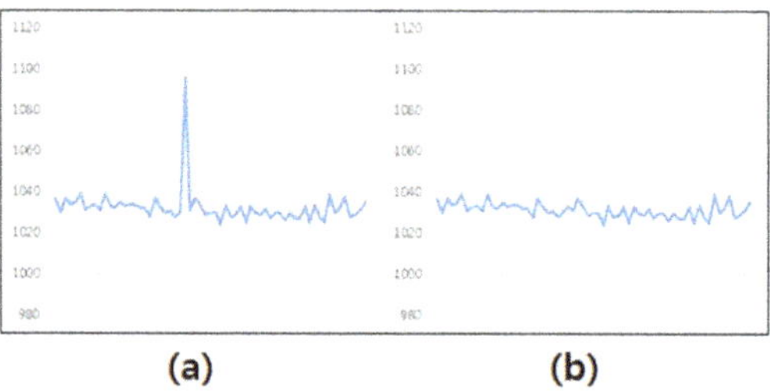

Figure 7. (**a**) Signal noise caused by heartbeat. (**b**) Results after the preprocessing.

3.2. Feature Extraction

Since signals acquired from the sensors differ according to the motion, as shown in Figure 5, we extract uncomplicated time series features to distinguish between gestures. Based on the results of the investigation, we selected two features which reflect the change and power of the signal.

The difference absolute mean value (DAMV) feature vector measures signal change equal to the average absolute difference of two sequential values as follows:

$$DAMV = \frac{\sum_{i=1}^{N-1} \left| X(i) - X((i+1)) \right|}{N-1} \tag{1}$$

The mean absolute value (MAV) is a measure of signal power which is equal to the average absolute value of the signal as follows:

$$MAV = \frac{\sum_{i=1}^{N} \left| X(i) \right|}{N} \tag{2}$$

3.3. Classification

We used an algorithm based on support vector machines (SVMs), which are well known to be the algorithm with the best generalization among machine algorithms. SVM is a supervised learning model widely used in classification and regression analysis. SVM maps features onto higher dimensions using a kernel function, and distinguishes them according to class, using hyperplanes. An appropriate SVM kernel must be selected to determine the decision boundaries between the different classes. We selected a radial basis function (RBF) kernel for non-linear classification [32].

$$k(x_i, x_j) = \exp(-\gamma \left\| x_i - x_j \right\|^2), \gamma > 0 \tag{3}$$

Here, γ is a kernel parameter which indicates the influence of squared Euclidean distance. We used the LIBSVM library, one of the most-used SVM libraries [32]. The two features mentioned in Section 4.2 were used as input. A classifier classifies inputs using a trained model generated by the training process.

In training, the variation of the signal is high when making a gesture, so it is presumed that the gesture is changed when the DAMV feature falls within the red circle, as shown in Figure 8. An SVM was trained using the MAV and DAMV features by monitoring the changes of signals. In each experiment, we acquired training data from each subject making each gesture for five seconds before the experiments began. To train the machine, we used five seconds of training data for each gesture from each subject. To investigate the optimal kernel parameter, we used several different pairs (C, γ) when the machine was trained, and selected an optimal parameter set empirically. In practice, there are limitations of a training process such as this in training the machine, but it guarantees certain training for specific subjects.

Since the period of feature extraction was about 30 Hz, the classifier was run using the trained model in real-time at 30 Hz. We asked subjects to make a gesture and produced a classification by confirming the concurrence of a classification result and the gesture the subject made.

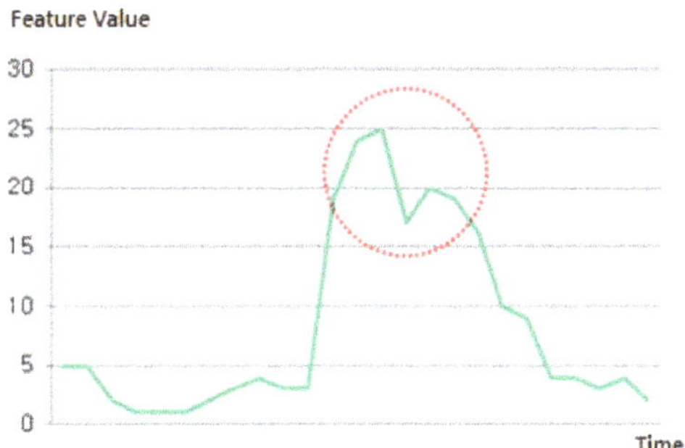

Figure 8. Example of extracted DAMV features. The circled area indicates when the subject made a gesture.

4. Experiments

In order to verify the performance of our proposed method, we conducted a comparison test between the proposed sensor, a commercial EMG sensor, and an FSRs sensor. We compared the accuracy of the proposed method with a commercial gesture recognition device and the results of previous research. To produce an objective comparison, we used the same experimental conditions, including the number of repetitions, gestures, and other factors, as used in previous studies. Lastly, we conducted an experiment using the gestures described above.

4.1. Comparison with EMG Sensor

The FETSA sensor was compared with an EMG sensor, which is the most intuitive and widely-used method for measuring muscle activity. A certified commercial EMG sensor, MyoWare Muscle Sensor of Advancer Technologies, was used. The sensor was attached above the extensor pollicis brevis muscle, which is responsible for the movement of the thumb, to detect the activities of the muscle. The FETSA sensor corresponds with channel 2 in Figure 4a. The sensor signals were recorded when the subject remained motionless and when the subject produced a "thumbs-up" motion.

Figure 9 shows the results of the comparison. In Figure 9a, signals were acquired when the subject remained motionless with the sensors attached, and in Figure 9b, signals were acquired while the subject produced the "thumbs-up" motion. As shown in Figure 9, the signal acquired from the FETSA was relatively uniform when the subject took no motion or made the "thumbs-up" motion. In contrast, in the EMG sensor, the signal had greater fluidity, even when the subject made no motion (Figure 9a). When the subject made the "thumbs-up" motion, the fluctuation in the signal was large, as it was affected by noise, and also static and dynamic contraction (Figure 9b). To analyze these results, the average and standard deviation of the signals were calculated. When the subjects made no motion, the average of the signal from FETSA was 970.08 and the standard deviation was 0.87, while the average of the signal from EMG was 68.90, with a standard deviation of 39.8. The average of the signal from FETSA was 995.18, and the standard deviation was 1.56. In contrast, the average of the signal from

EMG was 67.47, and the standard deviation was 67.46. These results indicate that the FETSA sensor is more robust to noise and more stable than the EMG sensor.

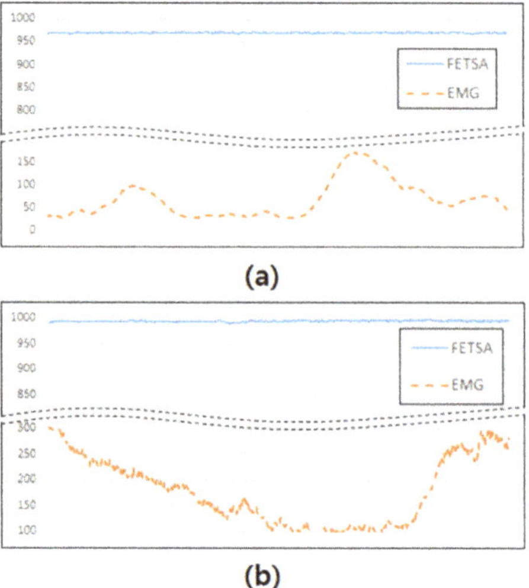

Figure 9. Comparison of the results using FETSA and EMG. (**a**) When the subject remained motionless. (**b**) When the subject produced the "thumbs-up" motion.

4.2. Comparison with the FSR Sensor

Although the robustness against noise of the FETSA sensor was demonstrated in Section 4.1 in comparison with the EMG sensor, it remained to be investigated whether FETSA can recognize gestures effectively. Therefore, we compared FETSA with the FSRs sensor, one of the most widely-used sensors for gesture recognition. A commercial FSRs sensor, RA18-DIY of Marveldex, was used. We attached the sensor over the same muscle that was used for the EMG and FETSA sensors in the previous section.

As shown in Figure 10, since the FSRs sensor is robust to electric noise, it acquired a more stable signal than the EMG sensor. When the subject was motionless, the average of the signal from the FETSA was 969.57, and standard deviation was 0.76. In case of FSRs, the average and standard deviation of the signal were 0.74 and 0.6 respectively. When the subject made the "thumbs-up" motion, the average of the signal from FETSA was 994.05, and the standard deviation was 1.26. In the case of the FSRs, the average and standard deviation of the signal were 2.98 and 1.01 respectively. Comparing both (a) and (b), the standard deviation of signals acquired from FETSA and FSRs sensors were similar, with a small fluctuation of 0.1–0.2. However, the sensors showed a difference in mean difference values. In the case of FETSA, the mean difference value was 26.48 between when the subject made the "thumbs-up" motion and when the subject remained motionless. In the FSRs, the mean difference value was 2.24. It is difficult to distinguish between the two conditions based on the signal from the FSR sensor, because the mean difference is low, with high standard deviation. However, it is easier to differentiate between the two conditions from the FETSA signals, which have a larger mean difference. The reason why the FSRs sensor does not have a high mean difference is that the signal does not increase when raising the thumb, due to the non-linearity of the FSRs sensor. Overall, these results indicate that the FETSA sensor is more effective than the FSRs sensor.

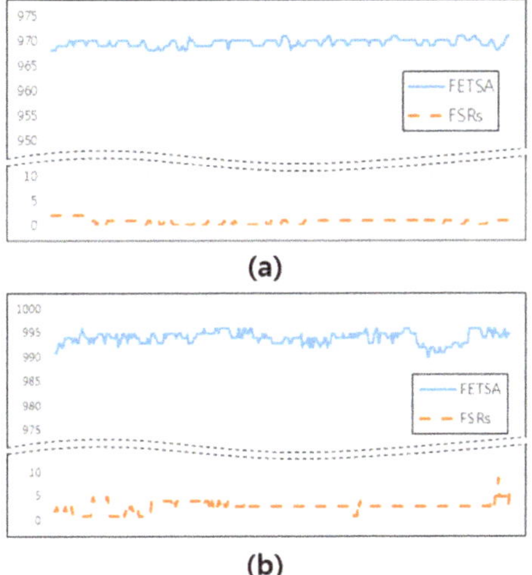

Figure 10. Comparison of the results using FETSA and FSRs: (**a**) the subject remained motionless, (**b**) the subjects made the "thumbs-up" motion.

4.3. Repeatability

Good repeatability is crucial for sensors, so we conducted an experiment to verify the repeatability of FETSA. A subject wearing the device was asked to clench and open his fist 10 times in a row.

Every channel of the sensor array was used in the repeatability test, and the results are shown in Figure 11. The same signal pattern was observed for each trial. To quantitative the results, the peak values of signals from the four channels were measured, and their averages and standard deviations were calculated. The average of the peak value was 1040.27 in channel 1, and the standard deviation was 0.93. The average of all channels was 998.94 and the standard deviation was 0.81. The standard deviations were very small compared to the average of peak values, indicating FETSA's ability to accurately measure repeated muscle activity.

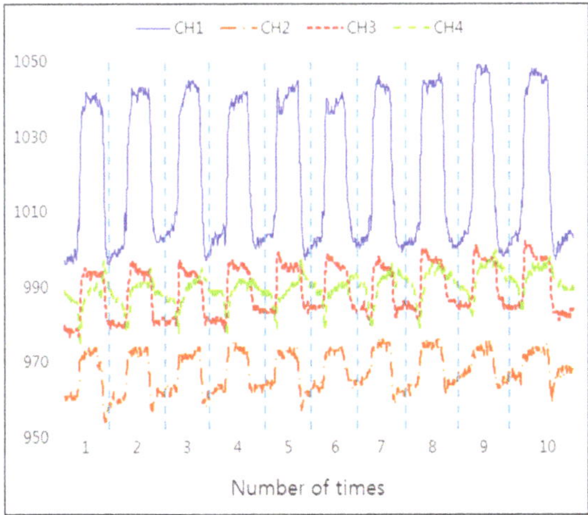

Figure 11. Results of repeatability tests.

4.4. Comparison with Contact Gesture Recognition Study

Pyeong-Gook Jung et al. [24] introduced a new method to detect muscular activity using air-pressure sensors. This approach overcomes the drawbacks of EMG and MMG sensors in detecting muscle activity and recognition of hand gestures. These researchers detected muscular activity by measuring the change in air pressure at air-pressure sensors contacted with the muscle of interest. They used fuzzy logic to determine gestures from the role of the muscles in each gesture.

To compare the performance of FETSA with that of the previous study, we used the six gestures defined in Jung's research (Figure 12). The test conditions were made as similar as possible, to ensure valid comparisons (Table 1).

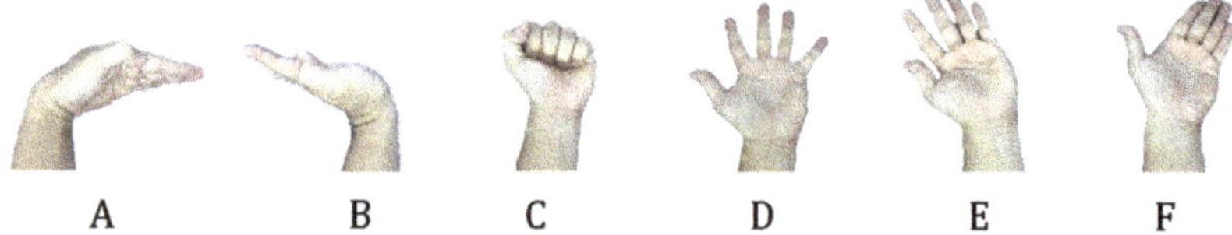

Figure 12. The six gestures that were defined in Jung's research.

Table 1. Comparison results for each subject.

Gesture	A	B	C	D	E	F
			Success(Proposed)/Success(previous)/Trial			
Subject A	**30**/30/30	**30**/30/30	**30**/29/30	**30**/29/30	**30**/28/30	**30**/30/30
Subject B	**18**/18/18	**19**/19/20	**22**/21/22	**16**/15/16	**20**/18/20	**15**/15/15
Subject C	**18**/17/18	**15**/14/15	**15**/15/15	**15**/14/15	**15**/15/15	**15**/16/17
Subject D	**20**/19/20	**14**/13/14	**16**/16/16	**18**/16/18	**20**/18/20	**15**/14/15
Subject E	**17**/16/17	**15**/16/17	**18**/17/18	**20**/18/20	**15**/14/15	**15**/14/15
Subject F	**18**/18/18	**15**/15/15	**16**/16/16	**16**/15/16	**17**/16/18	**20**/18/20
Total (%)	**100**/97.5/100	**97.3**/96.4/100	**100**/97.4/100	**100**/93.0/100	**99.1**/92.3/100	**98.2**/95.5/100

The average accuracy of FETSA was 99.1% while the average accuracy of the previous study was 95.35% [24]. FETSA was therefore more effective at determining the same gestures than the previous study.

4.5. Comparison with a Commercial Gesture Recognition Device

Myo, which is developed by Thalmic Labs Co., is a commercial gesture recognition device [5]. It measures EMG signals from sensors worn on the user's arm to control other digital devices. Myo provides five gestures (Figure 13).

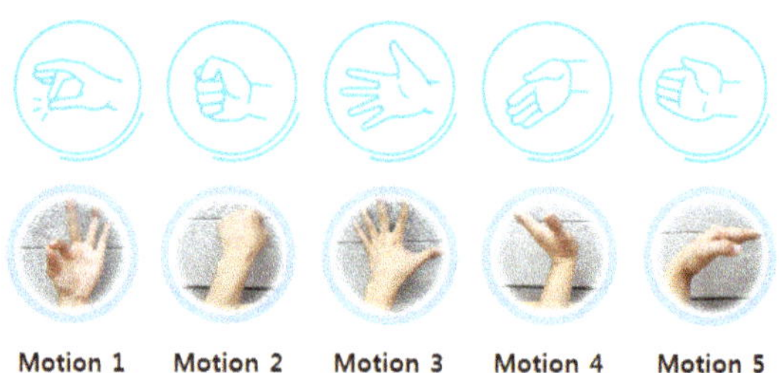

Figure 13. The five gestures provided by Myo.

We asked subjects to make the gestures while wearing the proposed device and Myo, and compared the accuracy of gesture recognition from the two devices. The subjects participating in the test were four men and four women. The results are presented in Table 2.

Table 2. Results of comparison tests between the proposed device and myo.

Gesture	Myo (Error Rate)	Proposed Device (Error Rate)
Motion 1	22.5	2.5
Motion 2	6.25	5
Motion 3	33.75	5
Motion 4	15	5
Motion 5	10	3.75
Total (%)	**17.5**	**4.25**

The average error rate of Myo was 17.5%. In contrast, the average error rate of FETSA was 4.25%. Myo recognized motion 3 as motion 2, and failed to recognize motion 1, resulting in a considerable increase in the average error rate. However, FETSA had a low average error rate and recognized the five gestures more accurately than Myo.

4.6. Hand Gesture Recognition with an FETSA Sensor

We performed a recognition experiment using the gestures defined in this research. The six hand gestures are shown in Figure 14: pinch of the finger ((1) in the Figure); flexion and extension of the fingers ((2) and (3)); flexion and extension of the wrist ((4) and (5)); extension of a thumb from a fist (6).

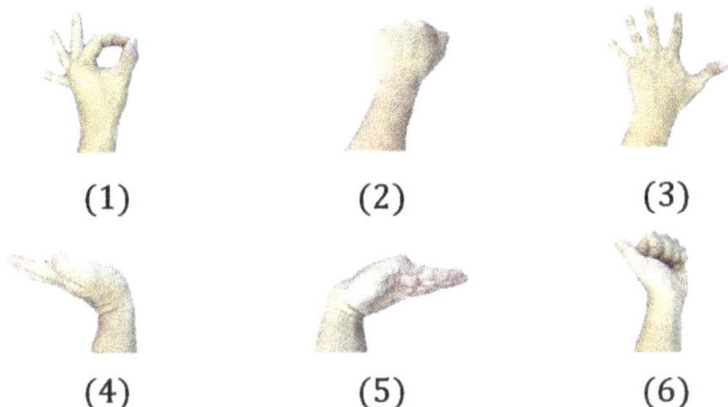

Figure 14. The six gestures defined in this research.

The eight subjects participating in this experiment were six men and two women. Before the experiment began, data from the subjects was used to train the SVM to recognize the gestures for five seconds. Each experiment was conducted 30 times per gesture, and the researcher randomly selected each gesture. The subjects made a gesture according to the researcher's instructions. The gesture recognition tests were repeated 1440 times. The results for the eight subjects are shown in Table 3. The average success rate for gesture recognition was 97.8%, and the number of misclassifications was very low at 2.2%.

Table 3. Comparison results for all subjects.

Gesture	1	2	3	4	5	6	Total (%)
1	234	0	0	2	0	2	98.3
2	0	235	1	0	3	2	97.5
3	0	0	239	1	1	0	99.2
4	1	2	0	234	1	1	97.9
5	0	3	0	2	232	0	97.9
6	5	0	0	1	3	235	96.3
Total (%)	97.5	97.9	99.5	97.5	96.6	97.9	**97.8**

5. Discussion and Conclusions

In this study, we developed a new gesture recognition method using FETSA, based on strain gauges to sense deformations. The sensor array was designed to overcome the drawbacks of other sensors and accurately recognize gestures. We fabricated a prototype clip-type device, providing enhanced usability in terms of wearing the sensor. Preprocessing algorithms were developed to remove noise from the acquired electrical signals. DAMV and MAV features were extracted from the signals, and gestures were recognized by an SVM, using the extracted features. The sensor array was shown to be able to analyze the eight motions of the wrist. We compared the performance of the sensor with those of a commercial EMG sensor and an FSRs sensor, which are commonly used in gesture recognition studies, under the same experimental conditions. We conducted additional experiments using the gestures defined in this research. As seen in the results, the proposed recognition method performed extremely well when compared with existing methods. However, it is difficult to directly compare our results with those of many other studies, due to the very different conditions involved, such as different types of gestures and different numbers of gestures.

Table 4 shows the results of previous gesture recognition studies which used a wide variety of techniques. Most recognition systems obtained accuracies of 80–90%, with an average accuracy of 90.93%. The results of this study, which produced 97–99% accuracy over the three experiments indicate that the proposed device is superior to those used in previous studies (Table 2). In future research we plan to study methods using both the movement and the location of a hand, combining the FETSA sensor with the IMU sensor.

Table 4. Results from previous studies on gesture recognition studies.

Sensor	Application	Algorithm	Accuracy
EMG & FSR [4]	Wrist	SVM	96%
EMG [33]	Finger	LDA	92%
Gyro sensor [1]	Hand, finger	-	98%
infrared sensor [34]	Wrist	Otsu's threshold	99%
OMTS [35]	Wrist	SVM	93%
EMG+IMU [36]	Wrist	LDA	96%
EMG+Inertial sensor [15]	Wrist	HMM	97.8%
EIT [37]	Wrist	SVM	90%
gyro sensor [38]	Wrist	-	96%
FSR [26]	Wrist	SVM	80%
EMG [2]	Wrist	HMM	89.60%
EMG [39]	Leg	LDA	90%
Flexible msg [40]	Glove	K-NN	93%
Gyro [41]	Hand	HMM	89%
EMG [18]	Brachial muscle	Fuzzy	92%
EMG [20]	Hand, Finger	HMM	90.5%
EMG [42]	Wrist	SVM	86%
MMG [43]	Forearm	LDA	89%
EMG [44]	Forearm, Finger	SVM	83%
EMG [45]	Finger	LDA	90%
MMG [28]	Brachial muscle	QDA	79.66%

Author Contributions: Conceptualization, S.-W.B.; methodology, S.-W.B. and S.-P.L.; investigation, S.-W.B.; writing—original draft preparation, S.-W.B.; writing—review and editing, S.-P.L.; project administration, S.-P.L.

Funding: This research received no external funding.

Conflicts of Interest: The authors declare no conflict of interest.

References

1. Xu, C.; Pathak, P.H.; Mohapatra, P. Finger-Writing with Smartwatch: A Case for Finger and Hand Gesture Recognition using Smartwatch. In Proceedings of the 16th International Workshop on Mobile Computing Systems and Applications, Santa Fe, NM, USA, 12–13 February 2015; pp. 9–14.
2. Lu, Z.; Chen, X.; Li, Q.; Zhang, X.; Zhou, P. A Hand Gesture Recognition Framework and Wearable Gesture Based interaction Prototype for Mobile Devices. *IEEE Trans. Human-Mach. Syst.* **2014**, *44*, 293–299. [CrossRef]
3. Chen, X.; Zhang, X.; Zhao, Z.; Yang, J.; Lantz, V.; Wang, K. Hand Gesture Recognition Research Based on Surface EMG Sensors and 2D-Accelerometers. In Proceedings of the 2007 11th IEEE International Symposium on Wearable Computers, Boston, MA, USA, 11–13 October 2007; pp. 11–14.
4. McIntosh, J.; McNeill, C.; Fraser, M.; Kerber, F.; Löchtefeld, M.; Krüger, A. EMPress: Practical Hand Gesture Classification with Wrist-Mounted EMG and Pressure Sensing. In Proceedings of the 2016 CHI Conference on Human Factors in Computing Systems, San Jose, CA, USA, 7–12 May 2016; pp. 2332–2342.
5. Sathiyanarayanan, M.; Rajan, S. MYO Armband for physiotherapy healthcare: A case study using gesture recognition application. In Proceedings of the 2016 8th International Conference on Communication Systems and Networks, Bangalore, India, 5–10 January 2016.
6. Lee, D.; Park, Y. Vision-based remote control system by motion detection and open finger counting. *IEEE Trans. Consum. Electron.* **2009**, *55*, 2308–2313. [CrossRef]
7. Lv, Z.; Halawani, A.; Feng, S.; Ur Réhman, S.; Li, H. Touch-Less Interactive Augmented Reality Game on Vision-Based Wearable Device. *Pers. Ubiquitous Comput.* **2015**, *19*, 551–567. [CrossRef]
8. Kim, Y.; Toomajian, B. Hand Gesture Recognition using Micro-Doppler Signatures with Convolutional Neural Network. *IEEE Access* **2016**, *4*, 7125–7130. [CrossRef]
9. Plouffe, G.; Cretu, A. Static and Dynamic Hand Gesture Recognition in Depth Data using Dynamic Time Warping. *IEEE Trans. Instrum. Meas.* **2016**, *65*, 305–316. [CrossRef]
10. Lu, W.; Tong, Z.; Chu, J. Dynamic Hand Gesture Recognition with Leap Motion Controller. *IEEE Signal Process. Lett.* **2016**, *23*, 1188–1192. [CrossRef]
11. Chaudhary, A.; Raheja, J.; Das, K.; Raheja, S. A Survey on Hand Gesture Recognition in Context of Soft Computing. In *Communications in Computer and Information Science, Proceedings of the International Conference on Computer Science and Information Technology, Bangalore, India, 2–4 January 2011*; Springer: Berlin/Heidelberg, Germany, 2011; pp. 46–55.
12. Raheja, J.; Minhas, M.; Prashanth, D.; Shah, T.; Chaudhary, A. Robust Gesture Recognition using Kinect: A Comparison between DTW and HMM. *Optik* **2015**, *126*, 1098–1104. [CrossRef]
13. Chaudhary, A.; Raheja, J. Light Invariant Real-Time Robust Hand Gesture Recognition. *Optik* **2018**, *159*, 283–294. [CrossRef]
14. Morganti, E.; Angelini, L.; Adami, A.; Lalanne, D.; Lorenzelli, L.; Mugellini, E. A Smart Watch with Embedded Sensors to Recognize Objects, Grasps and Forearm Gestures. *Procedia Eng.* **2012**, *41*, 1169–1175. [CrossRef]
15. Georgi, M.; Amma, C.; Schultz, T. Recognizing Hand and Finger Gestures with IMU Based Motion and EMG Based Muscle Activity Sensing. In Proceedings of the BIOSTEC 2015 the International Joint Conference on Biomedical Engineering Systems and Technologies, Lisbon, Portugal, 12–15 January 2015; pp. 99–108.
16. Wu, Y.; Chen, K.; Fu, C. Natural Gesture Modeling and Recognition Approach Based on Joint Movements and Arm Orientations. *IEEE Sens. J.* **2016**, *16*, 7753–7761. [CrossRef]
17. Wheeler, K.R.; Jorgensen, C.C. Gestures as Input: Neuroelectric Joysticks and Keyboards. *IEEE Pervasive Comput.* **2003**, *2*, 56–61. [CrossRef]
18. Khezri, M.; Jahed, M. A neuro–fuzzy Inference System for sEMG-Based Identification of Hand Motion Commands. *IEEE Trans. Ind. Electron.* **2011**, *58*, 1952–1960. [CrossRef]
19. Oonishi, Y.; Oh, S.; Hori, Y. A New Control Method for Power-Assisted Wheelchair Based on the Surface Myoelectric Signal. *IEEE Trans. Ind. Electron.* **2010**, *57*, 3191–3196. [CrossRef]

20. Liu, H. Exploring Human Hand Capabilities into Embedded Multifingered Object Manipulation. *IEEE Trans. Ind. Inf.* **2011**, *7*, 389–398. [CrossRef]
21. Kreil, M.; Ogris, G.; Lukowicz, P. Muscle activity evaluation using force sensitive resistors. In Proceedings of the 2008 5th International Summer School and Symposium on Medical Devices and Biosensors, Hong Kong, China, 1–3 June 2008.
22. Qi, J.; Jiang, G.; Li, G.; Sun, Y.; Tao, B. Intelligent Human-Computer Interaction Based on Surface EMG Gesture Recognition. *IEEE Access* **2019**, *7*, 61378–61387. [CrossRef]
23. Sun, X.; Yang, X.; Zhu, X.; Liu, H. Dual-Frequency Ultrasound Transducers for the Detection of Morphological Changes of Deep-Layered Muscles. *IEEE Sens. J.* **2018**, *18*, 1373–1383. [CrossRef]
24. Jung, P.; Lim, G.; Kim, S.; Kong, K. A Wearable Gesture Recognition Device for Detecting Muscular Activities Based on Air-Pressure Sensors. *IEEE Trans. Ind. Inf.* **2015**, *11*, 485–494. [CrossRef]
25. Lukowicz, P.; Hanser, F.; Szubski, C.; Schobersberger, W. Detecting and Interpreting Muscle Activity with Wearable Force Sensors. In *Pervasive Computing, Proceedings of the International Conference on Pervasive Computing, Dublin, Ireland, 7–10 May 2006*; Springer: Berlin/Heidelberg, Germany, 2006; pp. 101–116.
26. Dementyev, A.; Paradiso, J.A. WristFlex: Low-Power Gesture Input with Wrist-Worn Pressure Sensors. In Proceedings of the 27th Annual ACM Symposium on User Interface Software and Technology, Honolulu, HI, USA, 5–8 October 2014; pp. 161–166.
27. Comby, B.; Chevalier, G.; Bouchoucha, M. A New Method for the Measurement of Tremor at Rest. *Arch. Int. Physiol. Biochim. Biophys.* **1992**, *100*, 73–78. [CrossRef]
28. Zeng, Y.; Yang, Z.; Cao, W.; Xia, C. Hand-Motion Patterns Recognition Based on Mechanomyographic Signal Analysis. In Proceedings of the 2009 International Conference on Future BioMedical Information Engineering (FBIE), Sanya, China, 13–14 December 2009; pp. 21–24.
29. Courteville, A.; Gharbi, T.; Cornu, J. MMG Measurement: A High-Sensitivity Microphone-Based Sensor for Clinical use. *IEEE Trans. Biomed. Eng.* **1998**, *45*, 145–150. [CrossRef]
30. Watakabe, M.; Mita, K.; Akataki, K.; Itoh, Y. Mechanical Behaviour of Condenser Microphone in Mechanomyography. *Med. Biol. Eng. Comput.* **2001**, *39*, 195–201. [CrossRef]
31. Byun, S.W.; Lee, S.P. Hand Gesture Recognition Suitable for Wearable Devices using Flexible Epidermal Tactile Sensor Array. *J. Electr. Eng. Technol.* **2018**, *13*, 1731–1738.
32. Chang, C.C.; Lin, C.J. LIBSVM—A Library for Support Vector Machines. Available online: https://www.csie.ntu.edu.tw/~{}cjlin/libsvm/ (accessed on 17 August 2019).
33. Kim, J.; Cho, D.; Lee, K.J.; Lee, B. A Real-Time Pinch-to-Zoom Motion Detection by Means of a Surface EMG-Based Human-Computer Interface. *Sensors* **2014**, *15*, 394–407. [CrossRef] [PubMed]
34. Lee, D.; Son, Y.; Kim, B.; Kim, M.; Jeong, H.; Cho, I. Hand Gesture Segmentation Method using a Wrist-Worn Wearable Device. *J. Ergon. Soc. Korea* **2015**, *34*, 541–548. [CrossRef]
35. Luan, J.; Chien, T.; Lee, S.; Chou, P.H. HANDIO: A Wireless Hand Gesture Recognizer Based on Muscle-Tension and Inertial Sensing. In Proceedings of the 2015 IEEE Global Communications Conference (GLOBECOM), San Diego, CA, USA, 6–10 December 2015; pp. 1–7.
36. Huang, Y.; Guo, W.; Liu, J.; He, J.; Xia, H.; Sheng, X.; Wang, H.; Feng, X.; Shull, P.B. Preliminary Testing of a Hand Gesture Recognition Wristband Based on EMG and Inertial Sensor Fusion. In *Intelligent Robotics and Applications, Proceedings of the International Conference on Intelligent Robotics and Applications, Portsmouth, UK, 24–27 August 2015*; Springer: Cham/Heidelberg, Germany, 2015; pp. 359–367.
37. Zhang, Y.; Harrison, C. Tomo: Wearable, Low-Cost Electrical Impedance Tomography for Hand Gesture Recognition. In Proceedings of the 28th Annual ACM Symposium on User Interface Software & Technology, Charlotte, NC, USA, 11–15 November 2015; pp. 167–173.
38. Zhao, Y.; Pathak, P.H.; Xu, C.; Mohapatra, P. Finger and Hand Gesture Recognition using Smartwatch. In Proceedings of the 13th Annual International Conference on Mobile Systems, Applications, and Services, Florence, Italy, 18–22 May 2015; p. 471.
39. Zhang, X.; Liu, Y.; Zhang, F.; Ren, J.; Sun, Y.L.; Yang, Q.; Huang, H. On Design and Implementation of Neural-Machine Interface for Artificial Legs. *IEEE Trans. Ind. Inf.* **2012**, *8*, 418–429. [CrossRef] [PubMed]
40. Kumar, P.; Verma, J.; Prasad, S. Hand Data Glove: A Wearable Real-Time Device for Human-Computer Interaction. *Int. J. Adv. Sci. Technol.* **2012**, *43*, 15–26.

41. Kratz, L.; Morris, D.; Saponas, T.S. Making Gestural Input from Arm-Worn Inertial Sensors More Practical. In Proceedings of the SIGCHI Conference on Human Factors in Computing Systems, Austin, TX, USA, 5–10 May 2012; pp. 1747–1750.
42. Saponas, T.S.; Tan, D.S.; Morris, D.; Turner, J.; Landay, J.A. Making Muscle-Computer Interfaces More Practical. In Proceedings of the SIGCHI Conference on Human Factors in Computing Systems, Atlanta, GA, USA, 10–15 April 2010; pp. 851–854.
43. Alves, N.; Chau, T. Recognition of Forearm Muscle Activity by Continuous Classification of Multi-Site Mechanomyogram Signals. In Proceedings of the 2010 Annual International Conference of the IEEE Engineering in Medicine and Biology, Buenos Aires, Argentina, 31 August–4 September 2010; pp. 3531–3534.
44. Saponas, T.S.; Tan, D.S.; Morris, D.; Balakrishnan, R.; Turner, J.; Landay, J.A. Enabling always-Available Input with Muscle-Computer Interfaces. In Proceedings of the 22nd Annual ACM Symposium on User Interface Software and Technology, Victoria, BC, Canada, 4–7 October 2009; pp. 167–176.
45. Wojtczak, P.; Amaral, T.G.; Dias, O.P.; Wolczowski, A.; Kurzynski, M. Hand Movement Recognition Based on Biosignal Analysis. *Eng. Appl. Artif. Intell.* **2009**, *22*, 608–615. [CrossRef]

Article

Quasi Single Point Calibration Method for High-Speed Measurements of Resistive Sensors

Jesús A. Botín-Córdoba [1], Óscar Oballe-Peinado [1,2], José A. Sánchez-Durán [1,2] and José A. Hidalgo-López [1,*]

[1] Departamento de Electrónica, Universidad de Málaga, Andalucía Tech, Campus de Teatinos, 29071 Málaga, Spain; jesus.botin@uma.es (J.A.B.-C.); oballe@uma.es (Ó.O.-P.); jsd@uma.es (J.A.S.-D.)

[2] Instituto de Investigación Biomédica de Málaga (IBIMA), 29010 Málaga, Spain

* Correspondence: jahidalgo@uma.es; Tel.: +34-951-952-263

Received: 16 September 2019; Accepted: 29 September 2019; Published: 30 September 2019

Abstract: Direct interface circuits are a simple, inexpensive alternative for the digital conversion of a sensor reading, and in some of these circuits only passive calibration elements are required in order to carry out this conversion. In the case of resistive sensors, the most accurate methods of calibration, namely two-point calibration method (TPCM) and fast calibration methods I and II (FCMs I and II), require two calibration resistors to estimate the value of a sensor. However, although FCMs I and II considerably reduce the time necessary to estimate the value of the sensor, this may still be excessive in certain applications, such as when making repetitive readings of a sensor or readings of a large series of sensors. For these situations, this paper proposes a series of calibration methods that decrease the mean estimation time. Some of the proposed methods (quasi single-point calibration methods) are based on the TPCM, while others (fast quasi single-point calibration methods) make the most of the advantages of FCM. In general, the proposed methods significantly reduce estimation times in exchange for a small increase in errors. To validate the proposal, a circuit with a Xilinx XC3S50AN-4TQG144C FPGA has been designed and resistors in the range (267.56 Ω, 7464.5 Ω) have been measured. For 20 repetitive measurements, the proposed methods achieve time reductions of up to 61% with a relative error increase of only 0.1%.

Keywords: direct interface circuits; calibration methods; error analysis; resistive tactile sensor; time-based measurement

1. Introduction

Sensors interfacing with digital devices may be one of the most popular topics in electronics today. We can find a multitude of very different applications where this type of circuit plays a fundamental role, such as checking on patients [1], soil pore-water salinity sensors [2], monitoring composting processes [3], or specially piezoresistive tactile sensors [4]. Direct interface circuits (DICs) [5] are a series of circuits based on methods that use a small number of additional components to make the connection between a sensor and a programmable digital device (PDD). There are DICs for resistive sensors [6–14], capacitive [15–17] or inductive [18–22] sensors, and even DICs built to measure any of them [23].

In some cases, the DICs use, together with the sensor, only additional passive components. In other cases, the additional components may be a transistor or even a logic gate. Another important difference between the various types of DICs is the use of analogue elements that may be included in the PDDs. For example, we currently find microprocessors that can include analogue-to-digital converters (ADCs) or analogue comparators. With performances being equal, DICs that do not require additional active components or analogue elements within the PDDs will obviously be preferred. This type of DICs usually performs a time-to-digital conversion in the PDD in order to obtain the value of the physical magnitude to be measured and will be used in this paper for the specific case of resistive sensors. Using a DIC can result in a reduction in costs, complexity, and power consumption in the measurement chain compared to the use of a traditional scheme with ADCs and signal conditioning circuits.

DIC parameters such as uncertainties, effective numbers of bits, calibrations, resolution, or response to dynamic signals have already been well explored and analyzed [9,22,24–28]. However, the time needed to perform the conversion has received little attention when this parameter is in fact one of the most important characteristics for sensors, given the need for data acquisition speeds. This is crucial in the case of DICs for resistive sensors based on time-to-digital conversion since the sensor's resistance value range can be very broad, meaning an extensive time range is necessary for conversion. Furthermore, in order to improve accuracy in estimating the resistance value of the sensor, R, it is necessary to carry out additional readings of certain calibration resistors, therefore increasing the total time necessary to obtain R. This problem is increased when reading groups or arrays of resistive sensors, as usually happens in applications with tactile sensors or electronic skins.

The aim of the calibration resistors is to eliminate the influence that different parameters have on accuracy in estimating R. Apart from the internal resistance of the buffers of the pins of the PDDs, Ro, other factors that introduce errors in the measurement are the variability in: the capacitance used in the estimation process, C, the value in the supply voltage, V_{DD}, and the threshold voltage of the pins of the PDDs, V_f. All these possible sources of error can be compensated thanks to calibration resistors. Another possible source of error would be the self-heating of the resistors. However, this phenomenon can be neglected in DICs, since the current does not pass through the resistors continuously and its value is low due to the high resistance normally shown by resistive sensors.

Since calibration resistors must also increase their value in line with the increase in the maximum resistance value to be measured [26], the total time needed to estimate R, $T_E(R)$, can result in a high value. If the resistance value of a sensor is only to be measured sporadically, this problem may not be particularly important. However, if repetitive measurements are to be made for a single sensor or information can be obtained for different sensors (or even if these two situations occur simultaneously), then $T_E(R)$ becomes a fundamental parameter that can considerably slow down an application, or, in certain cases, even prevent the use of DICs. Finding a method to reduce $T_E(R)$ can be crucial in some practical applications, such as tactile sensors, where the number of resistive sensors to be read can be very high and, moreover, a high reading frequency of the sensors is required in order to calculate characteristics such as grips or slippages [29,30].

This paper presents a method for estimating R which allows $T_E(R)$ to be reduced without the need for additional hardware while maintaining adequate accuracy in the measurement, making it possible to use DICs in applications such as those indicated above.

The structure of the paper is as follows. In Section 2 different calibrations methods are evaluated regarding accuracy and time needed to estimate R. Section 3 presents two new calibration methods to increase the speed of a DIC. In Section 4, experimental results and discussion about these new methods are performed. Finally, Section 5 presents the main conclusions.

2. Evaluating *R* Estimation Time in Classical Calibration Methods

There are three classic calibration methods for a DIC, differing in the number of calibration resistors and how these resistors are used to discharge the capacitor. The circuits used in each of the methods are shown in Figure 1. Figure 1a shows the circuit for the single-point calibration method (SPCM). In this circuit, a capacitor is first charged to V_{DD} through the Pp pin of the PDD (resistor Rp is optional and only necessary if the current through the pin needs to be limited, although in the literature we can also find that this resistor can reduce the influence of the power-supply noise [31]). After charging, a discharge is made either through R or through a calibration resistor, R_{c0}. This is done by configuring the pin connected to the resistor that discharges as a logic 0 output, while the pin of the other resistor is configured as a high-impedance output. This way of proceeding in the discharge of a resistor will be called the normal discharge procedure. The discharge ends when the Pp pin (configured as the high-impedance input during discharge) detects a change to the logic 0 input. The capacitor is then re-charged and finally discharged again through the other resistor until a new logic 0 input is detected.

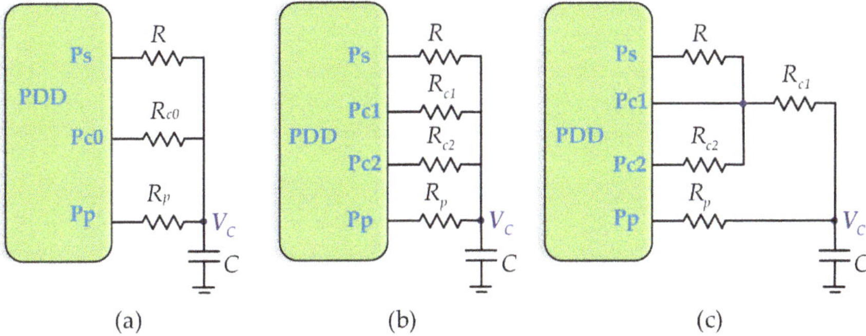

Figure 1. Different types of direct interface circuits (DICs): (**a**) single-point calibration method (SPCM); (**b**) two-point calibration method (TPCM); (**c**) three-signal calibration method (TSCM).

This method, taking into account the discharge equations of a capacitor, obtains [5]:

$$\frac{T_R}{T_{Rc0}} = \frac{(R+R_o)C\ln\frac{V_{DD}}{V_f}}{(R_{Rc0}+R_o)C\ln\frac{V_{DD}}{V_f}} = \frac{R+R_o}{R_{c0}+R_o} \quad (1)$$

where T_R and T_{Rc0} are the discharge times from V_{DD} to V_f through R and R_{c0}, respectively (the sub-index of the time measurement will always indicate which resistor it is discharged through). These times will be expressed as the number of PDD clock cycles occurring during the discharge process. Since R_o is unknown, we cannot obtain the value of R based on Equation (1). However, if, as usually occurs, R_o is a small value compared to those of the different resistors to be measured, it is possible to approximate:

$$R \approx \frac{T_R}{T_{Rc0}} R_{c0} \quad (2)$$

Improving estimation of R provided by Equation (2) and eliminating error due to R_o requires methods and circuits that use two calibration resistors: two-point calibration method (TPCM) and three-signal calibration method (TSCM) [5]. The TPCM, as shown in Figure 1b, carries out the same processes as the SPCM, but with two calibration resistors, R_{c1} and R_{c2}. In this method, the calculation of R is given [5] by:

$$R = \frac{T_R - T_{Rc1}}{T_{Rc2} - T_{Rc1}}(R_{c2} - R_{c1}) + R_{c1} \quad (3)$$

For its part, the TSCM, as shown in Figure 1c, performs three discharge processes, like the TPCM, but through R_{c1}, R+R_{c1}, and R_{c2}+R_{c1}. Proceeding in this way makes the equation for determining R simpler [32], which can be written as:

$$R = \frac{T_{R+Rc1} - T_{Rc1}}{T_{Rc2+Rc1} - T_{Rc1}} R_{c2} \quad (4)$$

$T_E(R)$ is different for each of the aforementioned methods. If we define T_{charge} as the time necessary for the capacitor charge, the values of $T_E(R)$ for the different methods are given by:

$$T_E(R, SPCM) = 2T_{\text{charge}} + T_R + T_{Rc0} \quad (5)$$

$$T_E(R, TPCM) = 3T_{\text{charge}} + T_R + T_{Rc1} + T_{Rc2} \tag{6}$$

$$T_E(R, TSCM) = 3T_{\text{charge}} + T_{R+Rc1} + T_{Rc2+Rc1} + T_{Rc1} \tag{7}$$

where the second term in parenthesis for T_E indicates which calibration method it is calculated with.

Two aspects must be considered with regard to Equations (5)–(7). Firstly, T_{charge} is much lower than the other times of these equations, since, as mentioned, R_p is either very small or not necessary, meaning we can eliminate this term from these equations. Secondly, calibration resistors are not the same in all methods. A common choice for the SPCM is to place R_{c0} in the middle of the range of resistors to be measured [R_{min}, R_{max}], in order to minimize the maximum error when using Equation (2). Given this, Equation (5) becomes:

$$T_E(R, SPCM) \approx T_R + T_{\frac{Rmax+Rmin}{2}} \approx T_R + \frac{T_{\text{Rmax}} + T_{\text{Rmin}}}{2} \tag{8}$$

where we have taken into consideration that R_o is very small compared to those of the resistors to be measured.

For its part, optimal performance in the TPCM requires R_{c1} and R_{c2} to be, respectively, in 15% and 85% of the resistance value range to be measured in order to minimize errors in estimation [26], meaning Equation (6) can be written as:

$$T_E(R, TPCM) \approx T_R + \underbrace{0.15 \cdot T_{\text{Rmax}} + 0.85 \cdot T_{\text{Rmin}}}_{T_{Rc1}} + \underbrace{0.85 \cdot T_{\text{Rmax}} + 0.15 \cdot T_{\text{Rmin}}}_{T_{Rc2}} = T_R + T_{\text{Rmax}} + T_{\text{Rmin}} \tag{9}$$

The literature does not describe any criteria for choosing R_{c1} and R_{c2} in the TSCM, and, although it is at least possible to maintain the 85% criterion for R_{c2}, it is obvious that R_{c1} cannot be in 15% of the range as its minimum value is R_{min} $+R_{c1}$. The systematic error made by this method will therefore always be greater than that obtained by the TPCM [33]. In any case, if we maintain the same criteria of 15% and 85% to situate the calibration resistors, we will have:

$$\begin{aligned} T_E(R, TSCM) &\approx T_R + \underbrace{0.15 \cdot T_{\text{Rmax}} + 0.85 \cdot T_{\text{Rmin}}}_{T_{R+Rc1}} + \underbrace{0.15 \cdot T_{\text{Rmax}} + 0.85 \cdot T_{\text{Rmin}}}_{T_{Rc1}} + \underbrace{0.85 \cdot T_{\text{Rmax}} + 0.15 \cdot T_{\text{Rmin}}}_{T_{Rc2}+T_{Rc1}} = \\ &= T_R + 1.15 \cdot T_{\text{Rmax}} + 1.85 \cdot T_{\text{Rmin}} \end{aligned} \tag{10}$$

Comparing Equations (9) and (10), it is obvious that $T_E(R,\text{TSCM}) > T_E(R,\text{TPCM})$, meaning that, in terms of temporal performances for the same accuracy, the TPCM outperforms the TSCM [33]. It is also obvious that $T_E(R,\text{SPCM}) < T_E(R,\text{TPCM})$, although the TPCM is more accurate and thus, the application for which the DIC is used will determine which one to use. Given the foregoing, the ideal situation would be to find a calibration method with a $T_E(R)$ similar to or lower than $T_E(R,\text{SPCM})$ and an accuracy equivalent to that obtained by the TPCM. The following section presents two new calibration methods that meet these two requirements when measuring the resistance value of a large number of sensors using the same DIC or carrying out repetitive measurements of the same sensor. We call these new methods quasi single-point calibration methods (QSPCMs).

3. Quasi Single-Point Calibration Methods

3.1. Quasi Single-Point Calibration Method

Let us suppose that, as in the TPCM, we want to use two calibration resistors in order to repetitively obtain the resistance value of a sensor (with the aim of achieving the maximum accuracy). Let T^n_R, T^n_{Rc1} and T^n_{Rc2} be the discharge times to obtain the nth estimation of R, R^n (the superscript will be used to indicate the estimation number in all variables). Since the calibration resistors are fixed, we have the following relationship if Equation (3) is used for an initial estimation of R:

$$\frac{T_R^0 - T_{Rc1}^0}{T_{Rc2}^0 - T_{Rc1}^0} = \frac{\left(R^0 + R_o^0\right)C\ln\frac{V_{VDD}^0}{V_f^0} - \left(R_{c1} + R_o^0\right)C\ln\frac{V_{VDD}^0}{V_f^0}}{\left(R_{c2} + R_o^0\right)C\ln\frac{V_{VDD}^0}{V_f^0} - \left(R_{c1} + R_o^0\right)C\ln\frac{V_{VDD}^0}{V_f^0}} = \frac{R^0 - R_{c1}}{R_{c2} - R_{c1}} \tag{11}$$

It should be remembered that in Equation (11) V_{DD}, V_f, and R are considered to be the same for the three measurements, since the temporal moments of the discharges of the series of measurements for an estimation are very close to each other (this approximation is one of the sources of error in any type of DIC). T_R^n, T_{Rc1}^n, and T_{Rc2}^n are measured again in any new estimation of R and R^n, thus updating the values of the voltages and R_o.

However, if the aim is to speed up the estimation of R^n, we can also use the data from the initial estimation of R as follows:

$$\begin{aligned}\frac{T_R^n - T_{Rc1}^n}{T_{Rc2}^0 - T_{Rc1}^0}\cdot\frac{T_{Rc1}^0}{T_{Rc1}^n} &= \frac{(R^n + R_o^n)C\ln\frac{V_{VDD}^n}{V_f^n} - (R_{c1} + R_o^n)C\ln\frac{V_{VDD}^n}{V_f^n}}{(R_{c2} + R_o^0)C\ln\frac{V_{VDD}^0}{V_f^0} - (R_{c1} + R_o^0)C\ln\frac{V_{VDD}^0}{V_f^0}}\cdot\frac{(R_{c1} + R_o^0)C\ln\frac{V_{VDD}^0}{V_f^0}}{(R_{c1} + R_o^n)C\ln\frac{V_{VDD}^n}{V_f^n}} = \\ &= \frac{R^n - R_{c1}}{R_{c2} - R_{c1}}\cdot\frac{R_{c1} + R_o^o}{R_{c1} + R_o^n}\end{aligned} \tag{12}$$

With Equation (12), we can solve R^n:

$$R^n = \frac{R_{c1} + R_o^n}{R_{c1} + R_o^0}\cdot\frac{T_R^n - T_{Rc1}^n}{T_{Rc2}^0 - T_{Rc1}^0}\cdot\frac{T_{Rc1}^0}{T_{Rc1}^n}\cdot(R_{c2} - R_{c1}) + R_{c1} \tag{13}$$

The first quotient of the member on the right of this equation is a term very close to one, since $R_o << R_{c1}$, and, moreover, the variations of R_o over time (due to the circuit conditions) are even smaller, meaning $R_o^0 \approx R_o^n$. Hence, if we define:

$$A = \frac{T_{Rc1}^0}{T_{Rc2}^0 - T_{Rc1}^0} \tag{14}$$

based on Equation (13), we can write Equation (15) that describes the QSPCM:

$$R^n = \frac{T_R^n - T_{Rc1}^n}{T_{Rc1}^n}A(R_{c2} - R_{c1}) + R_{c1} \tag{15}$$

According to the method determined by Equation (15), when estimating R^0 we will need to evaluate the discharges through R_{c1} and R_{c2}; for any other estimate of R, we only need to evaluate the discharge through R_{c1}. Equation (15) can be applied not only for a succession of measurements of the same sensor, but also when a series of resistive sensors is being measured (in this situation, R^n would be the nth resistive sensor of the series).

As the value of R_{c2} is close to the highest value to be measured and its discharge time is only necessary in evaluating R^0, it is significantly time-saving in the estimations of R^n. Moreover, it is obvious that the hardware necessary to use Equation (15) in a DIC is the same as for the TPCM, meaning there is no additional hardware cost. There may only be a computational cost in estimating R^0 derived from a quotient and an added multiplication in Equation (15) compared to with Equation (3). However, if the product $A(R_{c2} - R_{c1})$ is stored in the PDD memory, starting from R^1, the number of arithmetic operations is lower in the QSPCM than in the TPCM, since subtraction of the denominator of the quotient is removed.

The main drawback in using the QSPCM is the small increase in error due to use of the approximation that eliminates the first quotient of the term on the right side of Equation (13). This phenomenon is studied in Section 4.

The QSPCM allows for an additional reduction in the uncertainty of the estimation of R if a mean of the discharge times of the calibration resistors of the first j measurement cycles is used instead of T^0_{Rc1} and T^0_{Rc2}. This mean value, $\mu_j(T_{Rc})$, is defined by:

$$\mu_j(T_{Rc}) = \frac{1}{j}\sum_{i=0}^{j-1} T^i_{Rc} \tag{16}$$

where $1 < j << n$ is necessary in order to achieve significant time-savings in the measurements. We can also define a new A^j, similar to A that appears in Equation (14), taking these means into account:

$$A^j = \frac{\mu_j(T_{Rc1})}{\mu_j(T_{Rc2}) - \mu_j(T_{Rc1})} \tag{17}$$

Proceeding in this way, we obtain Equation (18) equivalent to Equation (15):

$$R^n = \frac{T^n_R - T^n_{Rc1}}{T^n_{Rc1}} A^j(R_{c2} - R_{c1}) + R_{c1},\ 1 < j \ll n \tag{18}$$

Equation (18) presents a lower uncertainty in estimating R and only a small increase in the temporal cost, thanks to the fact that $j << n$. Equation (18) defines a variant of the QSPCM that we will call QSPCM-j.

Obviously, QSPCM-j increases the total measurement time and the information to be stored in the PDD. However, in Section 4, we show how very low values of j are enough to improve uncertainty in the measurements and reduce the maximum errors.

3.2. Fast Quasi Single-Point Calibration Method

The total time needed to estimate R can still be reduced by applying the methods described in Reference [33], called fast calibration methods I and II (FCM I and FCM II). In essence, the FCM I aims to reduce the discharge time through the highest resistances. To achieve this, the discharge through R stops after a preset time T_x ($T_x > T_{Rc1}$), followed immediately by the discharge of the capacitor through R_{c1} until completion. This discharge procedure will be referred to as the R accelerates the discharge procedure. There is therefore a reduction in the measurement times of all resistors with $T_R > T_x$, and this reduction increases as the value of R increases. Using the FCM I, the value that T_R would have with the normal discharge procedure can be found, according to the expression:

$$T_R = \frac{T_{Rc1}}{T_{Rc1} - T'_{Rc1}(R)} T_x,\ T_R \geq T_x \tag{19}$$

where $T'_{Rc1}(R)$ is the time discharging through R_{c1} after discharging through R for time T_x. If we call the time used in the accelerated discharge procedure T^*_R, this can be calculated using the following expression:

$$T^*_R = T_x + T'_{Rc1}(R) \tag{20}$$

Let $\Delta T_R = \Delta T_R(R, R_{c1}, T_x)$ be the difference in measurement times between the normal discharge procedure and the accelerated discharge procedure for R. Its value is therefore given by:

$$\begin{aligned} \Delta T_R = \Delta T_R(R, R_{c1}, T_x) = T_R - T^*_R = T_R - \left(T_x + T'_{Rc1}(R)\right) = T_R - \left(T_x + \tfrac{T_R - T_x}{T_R} \cdot T_{Rc1}\right) = \\ = (T_R - T_x)\left(1 - \tfrac{T_{Rc1}}{T_R}\right) \end{aligned} \tag{21}$$

Equation (21) shows that, obviously, the reduction in measurement time increases as T_x decreases. However, the choice of T_x also has implications for the maximum error in estimating R. Thus, the smaller T_x (and also, in consequence, the time needed to find T_R), the greater the error in estimating R may be, although Reference [33] shows that there may be an optimal T_x values zone where this phenomenon does not occur.

For its part, in the FCM II, the accelerated discharge procedure applies to both R and R_{c2}, with $T_E(R)$ decreasing even more. This method will be used to reduce $T_E(R)$, even though it also comes at a small cost in terms of accuracy of results. Obviously, $T_{Rc1} < T_x < T_{Rc2}$ must occur in order to apply the method.

In the FCM II, using Equation (19) for both R and R_{c2}, the estimation of R is given by:

$$R = \begin{cases} \dfrac{T_R - T_{Rc1}}{\dfrac{T_{Rc1}}{T_{Rc1} - T'_{Rc1}(R_{c2})} T_x - T_{Rc1}} (R_{c2} - R_{c1}) + R_{c1}, & T_R \leq T_x;\ T_{Rc1} < T_x < T_{Rc2} \\ \dfrac{\dfrac{T_x}{T_{Rc1} - T'_{Rc1}(R)} - 1}{\dfrac{T_x}{T_{Rc1} - T'_{Rc1}(R_{c2})} - 1} (R_{c2} - R_{c1}) + R_{c1}, & T_R > T_x;\ T_{Rc1} < T_x < T_{Rc2} \end{cases} \tag{22}$$

Using T_R^n and T_{Rc2}^0 calculated from the measurements made by the accelerated discharge procedure, we obtain a new calibration method determined by the equations equivalent to Equations (14) and (15). This method, which we will call Fast-QSPCM, is defined by:

$$R^n = \begin{cases} \dfrac{T_R^n - T_{Rc1}^n}{T_{Rc1}^n} A^*(R_{c2} - R_{c1}) + R_{c1}, & T_R^n \leq T_x;\ T_{Rc1} < T_x < T_{Rc2} \\ \left(\dfrac{T_x}{T_{Rc1}^n - T_{Rc1}^{n'}(R^n)} - 1 \right) A^*(R_{c2} - R_{c1}) + R_{c1}, & T_R^n > T_x;\ T_{Rc1} < T_x < T_{Rc2} \end{cases} \tag{23}$$

where $T_{Rc1}^{n'}(R^n)$ is the time for which the capacitor is discharged through R_{c1} after having done so through R in the nth estimation. For its part, A^* is described as:

$$A^* = \frac{1}{\dfrac{T_x}{T_{Rc1}^0 - T_{Rc1}^{0'}(R_{c2})} - 1} \tag{24}$$

Operating in the same way with Equations (17) and (18), the equation that defines another calibration method, Fast-QSPCM-j, is obtained:

$$R^n = \begin{cases} \dfrac{T_R^n - T_{Rc1}^n}{T_{Rc1}^n} A^{*j}(R_{c2} - R_{c1}) + R_{c1}, & T_R^n \leq T_x;\ T_{Rc1} < T_x < T_{Rc2};\ 1 < j \ll n \\ \left(\dfrac{T_x}{T_{Rc1}^n - T_{Rc1}^{n'}(R^n)} - 1 \right) A^{*j}(R_{c2} - R_{c1}) + R_{c1}, & T_R^n > T_x;\ T_{Rc1} < T_x < T_{Rc2};\ 1 < j \ll n \end{cases} \tag{25}$$

where A^{*j} is the same as A defined in Equation (17) but is with T_{Rc2} of each estimation calculated from the measurements of the accelerated discharge procedure, similarly to that in Equation (24).

In order to compare $T_E(R)$ in the new methods to those obtained in Equations (8) and (9), we are going to use the mean values of $T_E(R)$ (R being constant) in estimating n resistive sensors (or n estimations of the same sensor), $\mu(T_E(R))$. In the case of the SPCM and the TPCM: $\mu(T_E(R), SPCM)$ and $\mu(T_E(R), TPCM)$ match the values of $T_E(R)$ in Equations (8) and (9), since $T_E(R)$ is the same in any estimation of R^n. However, in the QSPCM and in the Fast-QSPCM, the values of $T_E(R)$ differ in the first estimation. For its part, in QSPCM-j and Fast-QSPCM-j, the first j estimations have a different $T_E(R)$ value from those of the others. Table 1 shows the $\mu(T_E(R))$ for each method when n estimations of R are made.

Table 1. Mean time for an estimation of R, if n estimations are made, both in traditional methods and in the methods presented in this paper.

Method	Mean of $T_E R$ for n Estimations, $\mu(T_E(R))$
SPCM	$T_R + T_{Rc0}$
TPCM	$T_R + T_{R\max} + T_{R\min}$
Fast calibration method II (FCM II)	$\begin{cases} \mu(T_E(R, TPCM)) - \Delta T_R(R, R_{c1}, T_x), \ T_R > T_x \\ \mu(T_E(R, TPCM)), \ T_R \le T_x \end{cases}$
Quasi single-point calibration method (QSPCM)	$T_R + T_{Rc1} + \frac{T_{Rc2}}{n}$
Fast single-point calibration method (Fast-QSPCM)	$\begin{cases} \mu(T_E(R, QSPCM)) - \Delta T_R(R, R_{c1}, T_x) - \frac{\Delta T_R(R_{c2}, R_{c1}, T_x)}{n}, \ T_R > T_x \\ \mu(T_E(R, QSPCM)), \ T_R \le T_x \end{cases}$
QSPCM-j	$T_R + T_{Rc1} + j\frac{T_{Rc2}}{n}$
Fast-QSPCM-j	$\begin{cases} \mu(T_E(R, QSPCM - j)) - \Delta T_R(R, R_{c1}, T_x) - j\frac{\Delta T_R(R_{c2}, R_{c1}, T_x)}{n}, \ T_R > T_x \\ \mu(T_E(R, QSPCM - j)), \ T_R \le T_x \end{cases}$

4. Experimental Results and Discussion

The performances obtained with the QSPCMs, SPCM, and TPCM have been compared using an FPGA (Xilinx XC3S50AN-4TQG144C, Xilinx Inc., San Jose, CA, USA) [34] as a PDD mounted on a FR-4 fiberglass substrate with four layers. The FPGA uses a quartz crystal to generate a 50 MHz operating frequency and needs two regulators (TPS79912 and TPS79633, Texas Instruments, Dallas, TX, USA) to power the core of the device at 1.2 V and the I/O buffers at 3.3 V. This limits the noise that digital activity can generate on the device's input and output pins. The I/O buffers have been programmed to provide the maximum current allowed by the manufacturer (i.e., 24 mA) in order to maintain digital integrity of the signals. The resistance measurement results are transmitted via an SPI to a controller and finally sent to a PC via a USB flash drive.

A series of 20 resistors with values between 267.56 Ω and 7464.5 Ω was used in order to evaluate the performances of the different methods, as these values are within the range of a large number of resistive sensors and, in particular, tactile sensors. Selecting a 47 nF capacitor ensured the discharge times for all these resistors were measured using a 14-bit counter implemented in the FPGA, and, moreover, that the relative maximum error using any method never exceeded 3%. Besides, three additional resistors of 8170 Ω, 9056.1 Ω, and 9963.7 Ω were used solely for the Fast-QSPCM and Fast-QSPCM-j methods, since these methods allow the discharge time of these resistors to be measured with the 14-bit counter.

Apart from these resistors, R_{c0} = 3486.8 Ω was used as a calibration resistor for the SPCM, and R_{c1} = 1098.1 Ω and R_{c2} = 6165.3 Ω were used for the other methods. All the resistors were measured using an Agilent 34401A digital multimeter. A number of 500 measurement cycles were performed for each of the 23 resistors used in order to measure the maximum errors and uncertainty. These 500 cycles were repeated again each time the resistors are estimated using a different method. In each cycle, the discharge time was measured through the resistor to be estimated and through one or both calibration resistors, depending on the method used and the measurement cycle in question. Thus, for example, for the QSPCM, discharge is via R_{c2} only in the first measurement cycle, while in the remaining 499 cycles discharge is only through R_{c1} and R.

Figure 2a shows the maximum errors obtained for the SPCM, the TPCM, and the QSPCM, while Figure 2b shows these same errors but expressed as relative to the nominal value of the resistor to be measured. For easier viewing, it should be noted that the results are presented in a linear scale on the y-axis in Figure 2a, and in a log2 scale for Figure 2b. As expected, the biggest errors always occur in the SPCM, except in the vicinity of its calibration resistor, 3486.8 Ω. Only in this case, Equation (2) is a very good approximation to obtain R. The absolute error curve of this method shows a typical "V" shape. For their part, the TPCM and the QSPCM maintain very similar errors to each other throughout the entire range. Absolute errors are practically constant in the low resistance value zone, while absolute errors increase slowly, and relative errors remain practically constant for high resistance values. Especially striking is the low resistance values region, where the SPCM shows relative errors up to 6 times greater than those of the other methods (of which errors are practically identical).

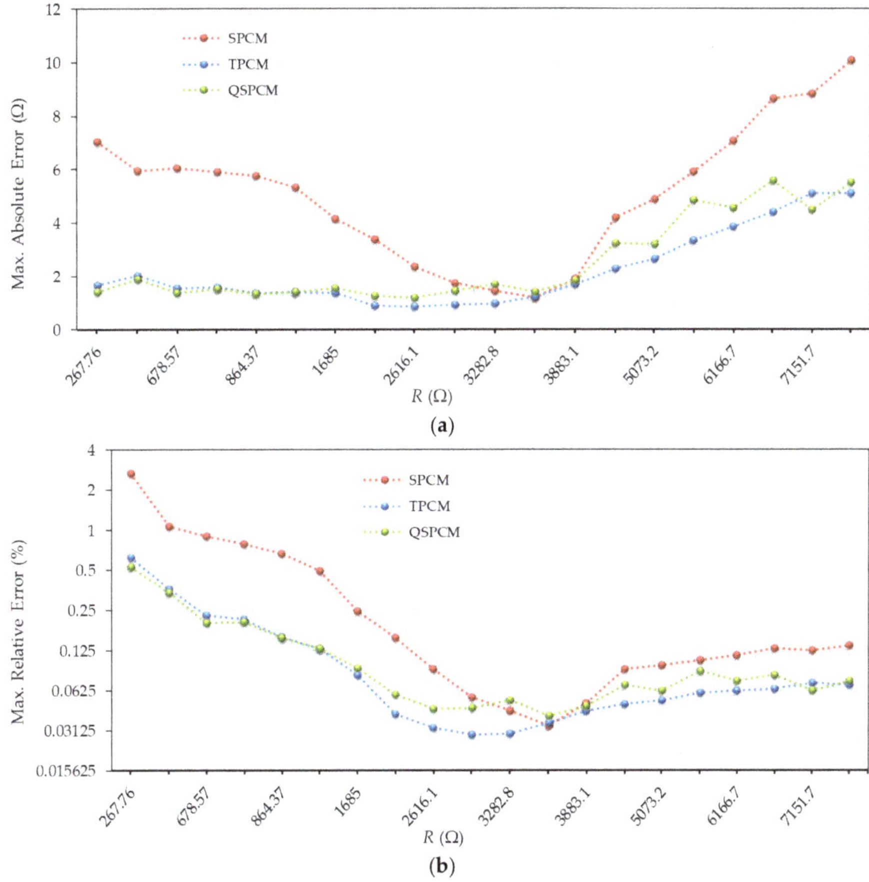

Figure 2. Errors in estimating resistance values using SPCM, TPCM, and QSPCM: (**a**) absolute maximum errors (linear scale); (**b**) relative maximum errors (log2 scale).

In order to evaluate the maximum errors in the Fast-QSPCM, Figure 3a compares these errors with those of the TPCM and FCM II. T_x = 163.84 µs was chosen for the comparison, i.e., half the time that can be measured with the 14-bit counter. This value was chosen as it is suitable for monitoring the value of the most significant bit of the counter, in order to know if time T_x was reached during the discharge, thus facilitating the hardware to be designed in the FPGA. This choice implies that all resistors with values under approximately 4000 Ω discharge the capacitor by themselves, while larger resistors use the accelerated discharge procedure to do so. However, as shown in Figure 3a, resistors with values below 4000 Ω do not present the same errors using TPCM and FCM II since, in this second method, R_{c2} is also evaluated using R_{c1}. Although this reduces $T_E(R)$, it also increases the error. For its part, the Fast-QSPCM in Figure 3a presents very similar errors to FCM II, and only shows to some degree greater errors than FCM II for some of the higher resistance values. It is again important to note that the TPCM can be used to measure a maximum resistance value of slightly more than 7500 Ω; however, thanks to the decrease in discharge times for large resistors, the FCM II and the Fast-QSPCM can be used to measure resistors of up to 10 kΩ. For this reason, the graphs in Figure 3 only show the results obtained with FCM II and Fast-QSPCM for resistors with values greater than 7500 Ω. Figure 3b shows the relative errors made by these methods, again in a log2 scale, where it is observed that the relative errors remain practically constant for large resistance values, with very close values in all three methods.

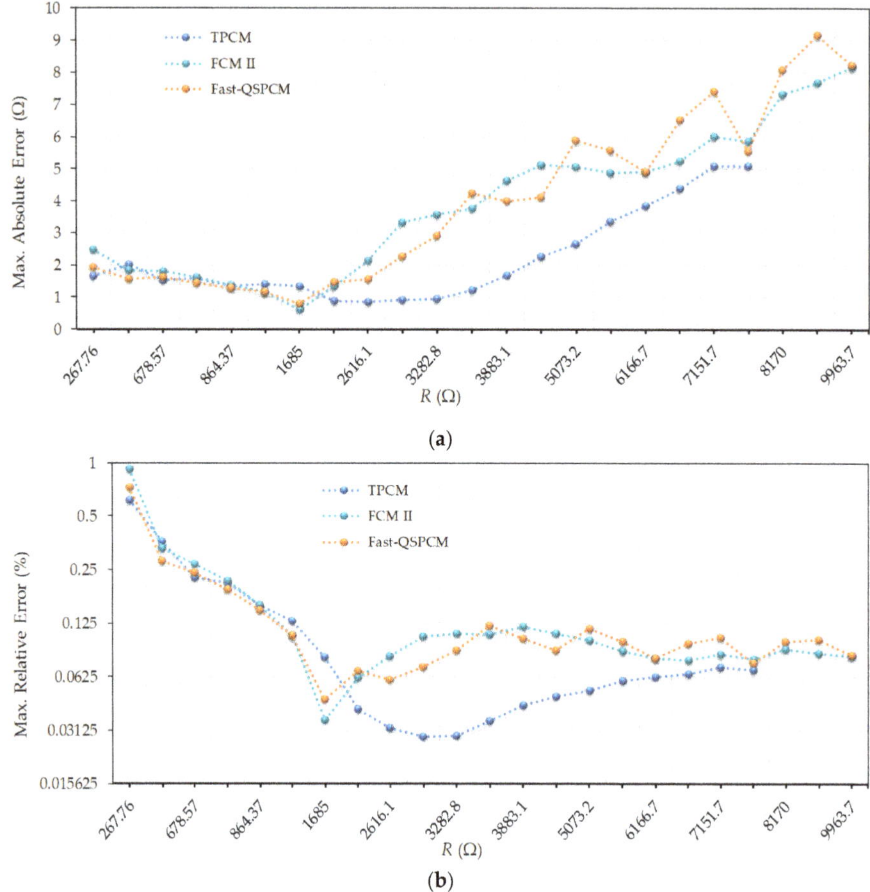

Figure 3. Errors in estimating resistance values using the TPCM, FCM II, and Fast-QSPCM: (**a**) absolute maximum errors (linear scale); (**b**) relative maximum errors (log2 scale).

Figure 4 compares the results obtained for QSPCM and QSPCM-j with $j \in \{2, 4, 16\}$. As expected, the shapes of the curves for the absolute (Figure 4a) and relative (Figure 4b) errors become gentler as j increases. However, there does not seem to be any truly significant improvements for $j > 2$, meaning $j = 2$ appears to be a good compromise in terms of error reduction, smoothness of the curve, and hardware complexity.

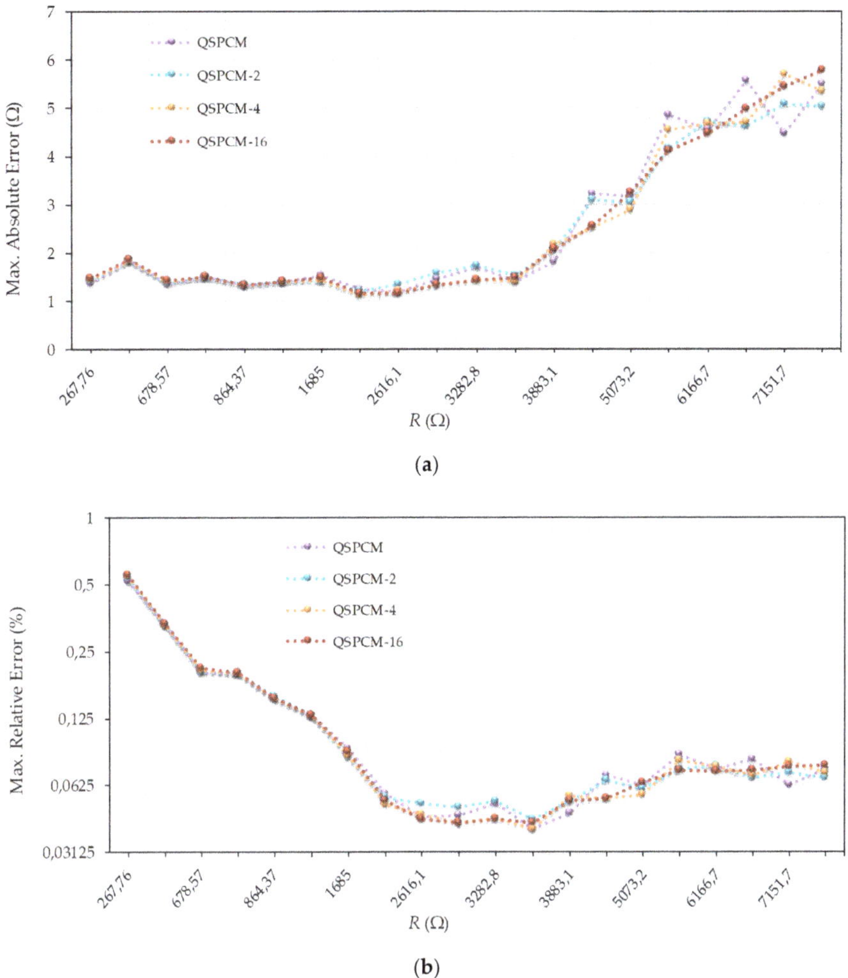

Figure 4. Comparison of errors made in estimating resistance values using the QSPCM and different values of j in QSPCM-j: (**a**) absolute maximum errors (linear scale); (**b**) relative maximum errors (log2 scale).

The results shown in Figure 5 are similar to the previous ones for Fast-QSPCM and Fast-QSPCM-j with $j \in \{2, 4, 16\}$. Again, an increase in j translates into smoother error curves for absolute (Figure 5a) and relative (Figure 5b) errors. However, these graphs show slightly better results than the rest for $j = 4$ and would seem to be the best choice. In fact, neither Figure 4 nor Figure 5 shows major differences between the results for the different methods. It is the designer who should select the most suitable method for each application.

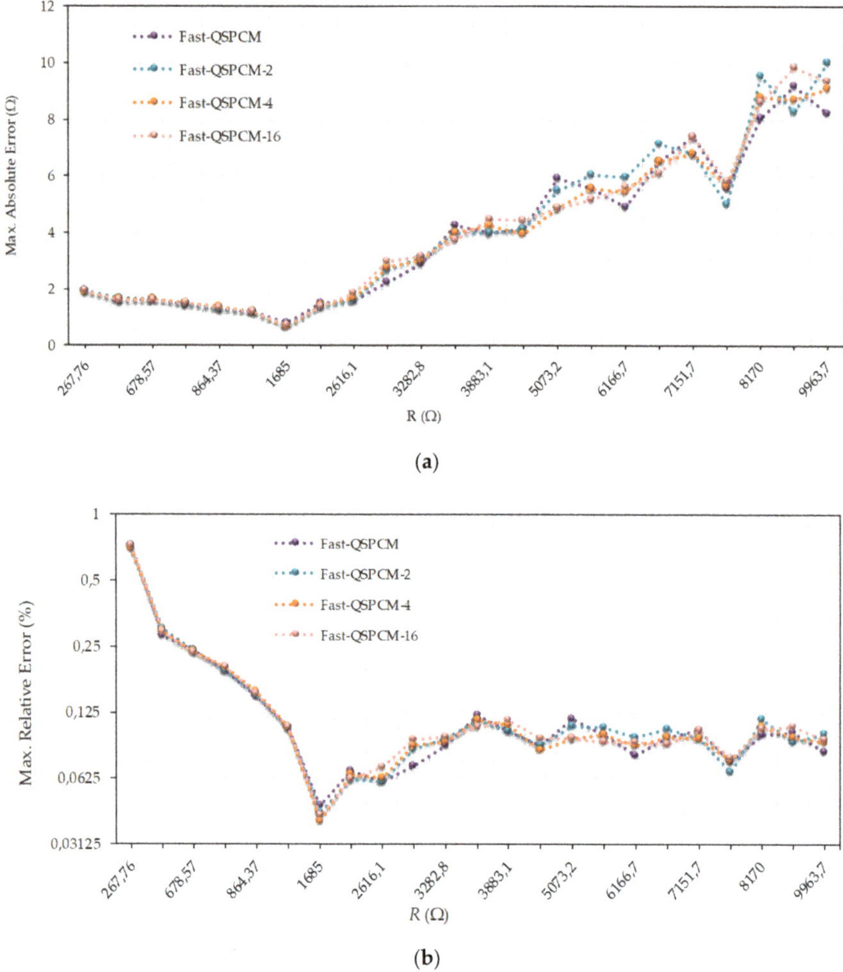

Figure 5. Comparison of errors made in estimating resistance values using Fast-QSPCM and different values of *j* in Fast-QSPCM-*j*: (**a**) absolute maximum errors (linear scale); (**b**) relative maximum errors (log2 scale).

To round off this section, Table 2 shows a summary of the performances of the methods analyzed when estimating a range of resistors with R_{max} = 7464.5 Ω and R_{min} = 267.56 Ω. The results are calculated based on data used for Figures 2–5 and equations listed in Table 1. The T_x = 163.84 µs was retained for the Fast-QSPCM and Fast-QSPCM-4 methods.

The second and third columns of Table 2 show the maximum errors obtained with any resistor within the range used. The penultimate column of the Table 2 shows the time needed to estimate R_{max} according to the different methods when estimating a single sensor once. For its part, the last column shows the case in which 20 estimations are made (for a single sensor or 20 different sensors). Particularly striking is the fact that the Fast-QSPCM reduces the mean time for 20 estimations by 61% compared to the TPCM, or by 47% compared to the SPCM. However, the maximum relative error only increases by 0.1% compared to that with the TPCM and is 3.65 times lower than that with the SPCM. For their part, the QSPCM and the QSPCM-2 present the lowest relative errors of all the methods, and yet show important reductions in the mean estimation times compared to the SPCM and the TPCM.

Table 2. Comparison of the performance of the different methods for a range of resistors with values between 267.56 Ω and 7464.5 Ω.

Method	Max. Absolute Error (Ω)	Max. Relative Error (%)	$\mu(T_E(R_{max}))$ µs 1 estimation	20 estimations
SPCM	10.07	2.63	449.83	449.83
TPCM	5.11	0.62	605.00	605.00
FCM II	6.00	0.92	409.11	409.11
QSPCM	5.56	0.52	605.00	364.54
Fast-QSPCM (T_x = 163.84 µs)	7.42	0.72	409.11	238.36
QSPCM-2	5.10	0.54	605.00	377.20
Fast-QSPCM-4 (T_x = 163.84 µs)	6.79	0.71	409.11	265.32

5. Conclusions

The TPCM is a calibration method commonly used in DICs for measuring resistive sensors, since it presents fewer errors in estimations than the SPCM. However, the TPCM needs more time to perform the estimation as it has to measure discharge times for three resistors: the resistor to be estimated and two calibration resistors.

Although the FCM has recently been proposed in order to reduce estimation time, this reduction may not be enough when having to make repetitive measurements of the values of a sensor or when having to measure the values of a large number of sensors. For these situations, this paper proposes a series of new methods, QSPCMs, in which two calibration resistors are also used but one of them is only measured in an initial estimation. Some of the proposed methods are based on the same discharge procedures as the TPCM, while others, in order to reduce the mean estimation time, use accelerated discharge procedures, as the FCM.

In order to compare the new proposed methods with traditional methods, a circuit with an FPGA (Xilinx XC3S50AN-4TQG144C) has been used to measure resistances values in the range (267.56 Ω, 7464.5 Ω). When 20 estimates of the range's maximum resistance value are made, one of the proposed methods, the Fast-QSPCM achieves a reduction of 61% in the mean estimation time compared to the TPCM, while the relative maximum error for any resistance value in the range only increases by 0.1%.

Author Contributions: Conceptualization, J.A.H.-L. and J.A.B.-C.; methodology, Ó.O.-P. and J.A.B.-C.; software, J.A.S.-D.; validation and investigation, J.A.H.-L., J.A.S.-D., J.A.B.-C., and Ó.O.-P.; writing of original draft preparation, J.A.B.-C.; writing of review and editing, J.A.H.-L. and Ó.O.-P.

Funding: This work was funded by the Spanish Government and by the European ERDF program funds under contract TEC2015-67642-R and by the Universidad de Málaga under program "Plan Propio 2018".

Conflicts of Interest: The authors declare no conflicts of interest.

References

1. Alhalabi, W. Patient monitoring at home using 32-channel cost-effective data acquisition device. *Telemat. Informa.* **2018**, *35*, 883–891. [CrossRef]
2. Skinner, A.J.; Lambert, M.F. An automatic soil pore-water salinity sensor based on a wetting-front detector. *IEEE Sens. J.* **2011**, *11*, 245–254. [CrossRef]
3. Rillo Moral, F. New autonomous sensor system for the continous monitoring of the composting. Ph.D. Thesis, Universitat Politècnica de Catalunya, Barcelona, Spain, 2016.
4. Hidalgo-López, J.A.; Oballe-Peinado, Ó.; Castellanos-Ramos, J.; Sánchez-Durán, J.A.; Fernández-Ramos, R.; Vidal-Verdú, F. High-accuracy readout electronics for piezoresistive tactile sensors. *Sensors* **2017**, *17*, 2513. [CrossRef]
5. Reverter, F.; Pallàs-Areny, R. *Direct Sensor To Microcontroller Interface Circuits*; Marcombo S.A.: Barcelona, Spain, 2005.
6. López-Lapeña, O.; Serrano-Finetti, E.; Casas, O. Low-power direct resistive sensor-to-microcontroller interfaces. *IEEE Trans. Instrum. Meas.* **2016**, *65*, 222–230. [CrossRef]

7. Dutta, L.; Hazarika, A.; Bhuyan, M. Direct interfacing circuit-based e-nose for gas classification and its uncertainty estimation. *IET Circuits, Devices Syst.* **2018**, *12*, 63–72. [CrossRef]
8. Reverter, F.; Casas, Ó. Interfacing differential resistive sensors to microcontrollers: A direct approach. *IEEE Trans. Instrum. Meas.* **2009**, *58*, 3405–3410. [CrossRef]
9. Kokolanski, Z.; Gavrovski, C.; Dimcev, V.; Makraduli, M. Hardware techniques for improving the calibration performance of direct resistive sensor-to-microcontroller interface. *Metrol. Meas. Syst.* **2013**, *20*, 529–542. [CrossRef]
10. Kokolanski, Z.; Gavrovski, C.; Dimcev, V. Improving the single point calibration technique in direct sensor-to-microcontroller interface. In Proceedings of the 19th Symposium IMEKO TC 4 Symposium and 17th IWADC Workshop, Advances in Instrumentation and Sensors Interoperability, Barcelona, Spain, 18–19 July 2013; pp. 87–91.
11. Reverter, F.; Horak, G.; Bilas, V.; Gasulla, M. Novel and low-cost temperature compensation technique for piezoresistive pressure sensors. In Proceedings of the XIX IMEKO World Congress, Fundamental and Applied Metrology, Lisboa, Portugal, 6–11 September 2009; pp. 2084–2087.
12. Dutta, L.; Hazarika, A.; Bhuyan, M. Nonlinearity compensation of DIC-based multi-sensor measurement. *Measurement* **2018**, *126*, 13–21. [CrossRef]
13. Sifuentes, E.; Casas, O.; Reverter, F.; Pallàs-Areny, R. Direct interface circuit to linearise resistive sensor bridges. *Sensors Actuators A Phys.* **2008**, *147*, 210–215. [CrossRef]
14. Sreenath, V.; Semeerali, K.; George, B. A resistance-to-digital converter possessing exceptional insensitivity to circuit parameters. In Proceedings of the 2016 IEEE International Instrumentation and Measurement Technology Conference Proceedings, Taipei, Taiwan, 23–26 May 2016.
15. Reverter, F.; Casas, Ò. Direct interface circuit for capacitive humidity sensors. *Sensors Actuators A Phys.* **2008**, *143*, 315–322. [CrossRef]
16. Gaitan-Pitre, J.E.; Gasulla, M.; Pallas-Areny, R. Direct interface for capacitive sensors based on the charge transfer method. In Proceedings of the 2007 IEEE Instrumentation and Measurement Technology Conference IMTC 2007, Warsaw, Poland, 1–3 May 2007.
17. Kyriakis-Bitzaros, E.D.; Stathopoulos, N.A.; Pavlos, S.; Goustouridis, D.; Chatzandroulis, S. A reconfigurable multichannel capacitive sensor array interface. *IEEE Trans. Instrum. Meas.* **2011**, *60*, 3214–3221. [CrossRef]
18. Ramadoss, N.; George, B. A simple microcontroller based digitizer for differential inductive sensors. In Proceedings of the 2015 IEEE International Instrumentation and Measurement Technology Conference Proceedings, Pisa, Italy, 11–14 May 2015.
19. Kokolanski, Z.; Jordana, J.; Gasulla, M.; Dimcev, V.; Reverter, F. Direct inductive sensor-to-microcontroller interface circuit. *Sensors Actuators A Phys.* **2015**, *224*, 185–191. [CrossRef]
20. Berine, P. Instrumentation and measurement technology and applications. *IEEE Instrum. Meas. Mag.* **2005**, *1*, 39. [CrossRef]
21. Asif, A.; Ali, A.; Abdin, M.Z.U. Resolution Enhancement in directly interfaced system for inductive sensors. *IEEE Trans. Instrum. Meas.* **2018**, *68*, 4104–4111. [CrossRef]
22. Sifuentes, E.; Gonzalez-Landaeta, R.; Cota-Ruiz, J.; Reverter, F. Measuring dynamic signals with direct sensor-to-microcontroller interfaces applied to a magnetoresistive sensor. *Sensors* **2017**, *17*, 1150. [CrossRef] [PubMed]
23. Czaja, Z. Time-domain measurement methods for R, L and C sensors based on a versatile direct sensor-to-microcontroller interface circuit. *Sensors Actuators A Phys.* **2018**, *274*, 199–210. [CrossRef]
24. Reverter, F.; Pallàs-Areny, R. Effective number of resolution bits in direct sensor-to-microcontroller interfaces. *Meas. Sci. Technol.* **2004**, *15*, 2157–2162. [CrossRef]
25. Custodio, A.; Pallas-Areny, R.; Bragos, R. Error analysis and reduction for a simple sensor-microcontroller interface. *IEEE Trans. Instrum. Meas.* **2001**, *50*, 1644–1647. [CrossRef]
26. Pallàs-Areny, R.; Jordana, J.; Casas, Ó. Optimal two-point static calibration of measurement systems with quadratic response. *Rev. Sci. Instrum.* **2004**, *75*, 5106–5111. [CrossRef]
27. Oballe-Peinado, Ó.; Vidal-Verdú, F.; Sánchez-Durán, J.A.; Castellanos-Ramos, J.; Hidalgo-López, J.A. Accuracy and resolution analysis of a direct resistive sensor array to FPGA interface. *Sensors* **2016**, *16*, 181. [CrossRef]
28. Nagarajan, P.R.; George, B.; Kumar, V.J. Improved single-element resistive sensor-to-microcontroller interface. *IEEE Trans. Instrum. Meas.* **2017**, *66*, 2736–2744. [CrossRef]

29. Oballe-Peinado, O.; Hidalgo-Lopez, J.A.; Castellanos-Ramos, J.; Sanchez-Duran, J.A.; Navas-Gonzalez, R.; Herran, J.; Vidal-Verdu, F. FPGA-based tactile sensor suite electronics for real-time embedded processing. *IEEE Trans. Ind. Electron.* **2017**, *64*, 9657–9665. [CrossRef]
30. Sánchez-Durán, J.A.; Hidalgo-López, J.A.; Castellanos-Ramos, J.; Oballe-Peinado, Ó.; Vidal-Verdú, F. Influence of errors in tactile sensors on some high level parameters used for manipulation with robotic hands. *Sensors* **2015**, *15*, 20409. [CrossRef] [PubMed]
31. Reverter, F.; Gasulla, M.; Pallàs-Areny, R. Analysis of power-supply interference effects on direct sensor-to-microcontroller interfaces. *IEEE Trans. Instrum. Meas.* **2007**, *56*, 171–177. [CrossRef]
32. Van Der Goes, F.M.L.; Meijer, G.C.M. A novel low-cost capacitive-sensor interface. *IEEE Trans. Instrum. Meas.* **1996**, *45*, 536–540. [CrossRef]
33. Hidalgo-López, J.A.; Botín-Córdoba, J.A.; Sánchez-Durán, J.A.; Oballe-Peinado, Ó. Fast calibration methods for resistive sensor readout based on direct interface circuits. *Sensors* **2019**, *19*, 3871. [CrossRef]
34. Xilinx, Inc. Spartan-3AN FPGA Family Data Sheet. Available online: http://www.xilinx.com/support/documentation/data_sheets/ds557.pdf (accessed on 29 September 2019).

Article

Assessment of Stickiness with Pressure Distribution Sensor Using Offset Magnetic Force

Takayuki Kameoka [1,*], Akifumi Takahashi [1], Vibol Yem [1], Hiroyuki Kajimoto [1], Kohei Matsumori [2], Naoki Saito [2] and Naomi Arakawa [2]

1 Department of informatics, The University of Electro-Communications, Tokyo 182-8585, Japan; a.takahashi@kaji-lab.jp (A.T.); yem@kaji-lab.jp (V.Y.); kajimoto@kaji-lab.jp (H.K.)

2 Shiseido Global Innovation Center, Kanagawa 220-0011, Japan; kohei.matsumori@shiseido.com (K.M.); naoki.saito@shiseido.com (N.S.); naomi.arakawa1@shiseido.com (N.A.)

* Correspondence: kameoka@kaji-lab.jp; Tel.: +81-42-443-5445

Received: 29 August 2019; Accepted: 27 September 2019; Published: 27 September 2019

Abstract: The quantification of stickiness experienced upon touching a sticky or adhesive substance has attracted intense research attention, particularly for application to haptics, virtual reality, and human–computer interactions. Here, we develop and evaluate a device that quantifies the feeling of stickiness experienced upon touching an adhesive substance. Keeping in mind that a typical pressure distribution sensor can only measure a pressing force, but not a tensile force, in our setup, we apply an offset pressure to a pressure distribution sensor and measure the tensile force generated by an adhesive substance as the difference from the offset pressure. We propose a method of using a magnetic force to generate the offset pressure and develop a measuring device using a magnet that attracts magnetic pin arrays and pin magnets; the feasibility of the method is verified with a first prototype. We develop a second prototype that overcomes the noise problems of the first, arising from the misalignment of the pins owing to the bending of the magnetic force lines at the sensor edges. We also obtain measurement results for actual samples and standard viscosity liquids. Our findings indicate the feasibility of our setup as a suitable device for measuring stickiness.

Keywords: haptics; measurement techniques; stickiness; sticky feeling

---

1. Introduction

Haptic perception has begun garnering increasing attention over the past few years, and in this regard, several studies have examined the representation of the human skin sensation, particularly in the fields of virtual reality and human–computer interactions. In reproducing a realistic feeling, it has been found effective to measure changes in the skin condition, such as the skin deformation distribution and contact area, in real-world situations and to reproduce this information. For example, Levesque et al. [1] measured the horizontal displacement of the skin of a finger tracing a glass surface in detail to capture information on how the skin deforms when in contact with an irregular shape. Bicchi et al. [2] captured changes in the skin contact area for a finger touching a flexible object. Such measurements are closely related to the technique used in tactile presentation [3]. The measurement of the horizontal displacement of skin is related to the development of devices that present horizontal displacement [4,5] and the measurement of the skin contact area has led to the development of devices that can represent a feeling of flexing/flexibility via changing the contact area [6,7].

Against this backdrop, here, we focus on the distribution of skin deformation corresponding to the feeling of "stickiness". In this study, stickiness is defined as the feeling experienced when touching an adhesive material such as glue, Nattō (which is a traditional Japanese food made from soybeans), or honey. The feeling of sticky sensation on the surface is also expressed as a frictional resistance [8]. Chen et al. [9]. investigated the correspondence between the measurement of physical properties of texture surface and subjective evaluation of touch sensation, and also mentioned the stickiness. However, we address stickiness that is related to the motion of the finger along surface normal, that is, the sensation experienced after releasing a finger that has been pressed against a sticky material. Here, we note that stickiness affects our impression of daily consumables such as lotion, creams, and so on. Moreover, stickiness is known as one of the factors responsible for the feeling of "wetness" in fabric perception [10,11]. Stickiness is often used as a general aversion sensation [12] and is a quality that is attributed to a wide range of materials and products.

Though Liu et al. measured adhesive force by the MEMS device [13], the development of the system corresponding to the adhesive force measurement on the skin is necessary in order to evaluate the sense of human. It has been speculated that both proprioceptive and cutaneous sensations contribute to stickiness; here, however, we mainly focus on cutaneous sensation. In this context, Yamaoka et al. [14] derived the relationship between the contact area of an adhesive surface and the temporal change in the pressing force, and found that there is large hysteresis in the contact area. The authors further created a stickiness display based on this finding. However, because their observations were limited to the change in the contact area, the detailed force distribution during the period of stickiness was not clearly elucidated. Such detailed physical properties are often required to be measured in the field of food [15,16]. For example, Dan et al. [17]. measured the bite force applied to raw and cooked apples using a pressure distribution sensor sheet.

In previous reports, we have described the basic principle of a system that measures the force distribution between adhesive substances and finger skin [18,19]. We used a pressure distribution sensor sheet to measure the adhesive force in the form of pressure distribution. Here, it must be noted that common pressure distribution sensors can measure only positive (compressive) pressure, but not negative pressure. Therefore, we devised a method of inserting a pin matrix between the skin and the pressure distribution sensor to apply a "preload" using the weight of the pins. With this configuration, the adhesive force can be observed as a decrease in the offset force when the finger is raised. In our previous studies, the weight of the pin was set to 0.8 g, and in the case of a highly sticky specimen, the pin could float because of the stronger adhesive force. Therefore, it was necessary to apply a stronger preload to the sensor to perform more stable measurements. The other issue in our previous studies involved the sensor sheet. Pressure-sensitive rubber sensors are prone to undesired current pathways and large hysteresis, both of which make it difficult to realize accurate sensing using prototype systems.

In this paper, we present and evaluate a stickiness measurement device with a large measurement range. When compared with our previous approaches, here, we used a load cell substrate with independent sensing points and applied a more powerful preload using a magnetic force.

2. Measuring System

2.1. Pressure Distribution Sensor Using Load Cell

To measure a distributed adhesive force, we developed a pressure distribution sensor using load cells. Figure 1 shows the load cell, schematic, and photo of sensor substrate. In our study, we used HSFPAR003A load cells (Alps, Inc. Tokyo, Japan) to measure the pressure; this load cell allows the measurement of forces up to 3.5 N. Sensing points are located at 2.54 mm intervals, which correspond to the two-point discrimination threshold of the human fingertip [20]; this interval is thus sufficient for measurement. One-unit board has 16 load cells. One load cell is selected and amplified by analog multiplexer (ADG726BCPZ, Analog Devices, Norwood, MA, USA) and differential amplifier (AD623ARMZ, Analog Devices, Norwood, MA, USA).

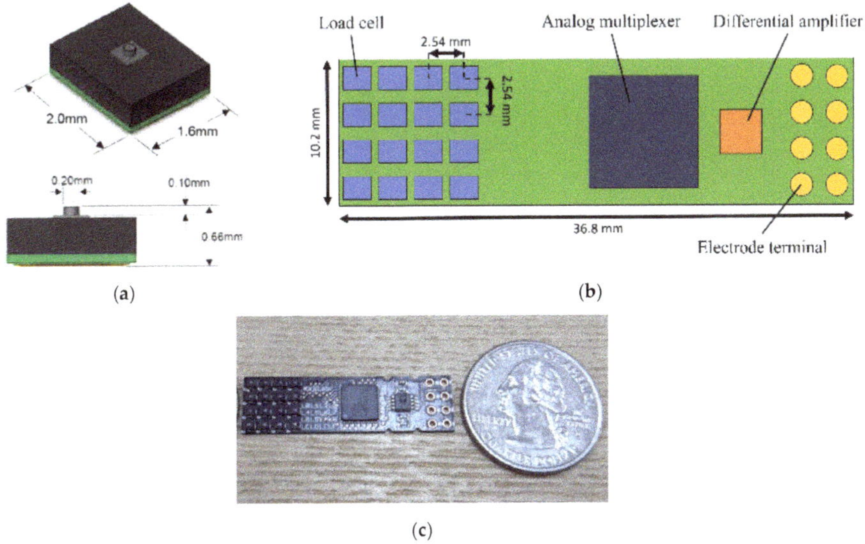

(c)

Figure 1. (**a**) Load cell used for pressure sensing. (**b**,**c**) Schematic and photo of sensor substrate comprising 4-by-4 load cells.

2.2. Offset Pressure Generated by Magnetic Force

Figure 2 shows the overall measurement system. The device consists of an acrylic pin insertion plate, an 8 × 12 magnetic pin array, three base magnets (50 mm × 50 mm × 10 mm), six load cell substrates (each mounted with 4 × 4 load cells), and an acrylic fixed pedestal.

In this study, to apply an offset preload to the load cell, a magnet was installed under the load cell substrate and a pin matrix made of pin magnets was inserted between the point of skin contact and the load cell substrate (Figure 2). Each pin of the magnet pin array was aligned to correspond to a single load cell. With this configuration, because the magnet under the substrate and the pin magnet matrix attract each other, an offset pressure (i.e., preload) can be applied to the load cell. Three magnets were stacked to strengthen the magnetic field (which was about 430 mT at the center). The magnetic field of adjacent pins might interfere with each other. However, because the polarities of the pin magnets are all set in the same direction, and the pin magnets are all at the same height and approximately horizontal, the force generated by the interference is the repulsive force in the tangential direction, which, in principle, has no effect on the normal force measurement. The load cells and magnet pin array were both positioned at 2.54 mm intervals. The pin magnet was 2 mm in diameter and 10 mm in height, and the magnetic force was 275 mT. The acrylic plate and the pedestal were made with a laser cutter. The acrylic plate was chosen because it is easy to cut using the laser cutter. The cut surface becomes slightly conical, and the contact area with the pin becomes smaller, reducing the friction.

We used six load cell substrates mounted on the base board, constituting 96 measurement points. The voltage signal from a single load cell is selected by the analog multiplexer, amplified by the differential amplifier, and then measured by an AD converter (MCP3208, Microchip Technology Inc., Chandler, AZ, USA). All operation is conducted by a micro-controller (mbed LPC1768, NXP Semiconductors N.V., Eindhoven, Netherland), and the data are sent to PC via USB serial port. Measurements of all points were conducted 60 times per second, and simple moving averages were calculated for noise removal (window size was 16).

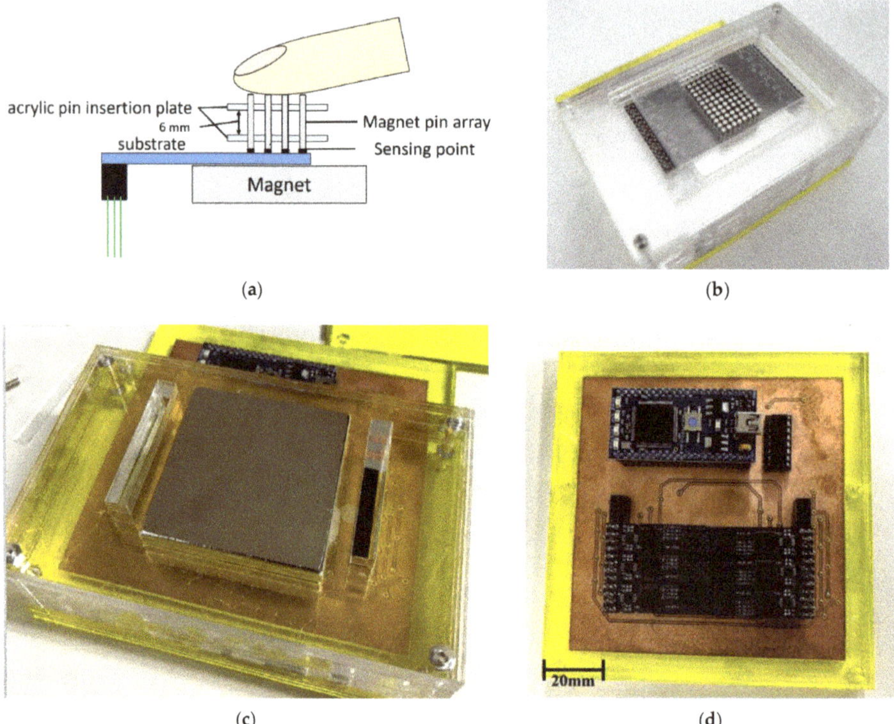

(a) (b)

(c) (d)

Figure 2. Measurement system based on magnet (**a**) structure, (**b**) oblique view, (**c**) base magnet, and (**d**) circuit board of sensing device. The attractive force between the base magnet and the pin magnets works as a preload to the load cell at the sensing point, enabling measurement of tensile force distribution of adhesive material. Six load cell substrates are mounted on the base board, constituting 96 measurement points.

We calibrated the load cell by adding a known weight on load cell. Figure 3 shows the result of single unit, showing high linearity ($R^2 > 0.99$) and the load cell value was 10.5 per 1 g. That is, the measured weight is 0.095 gf (=0.94 mN) per value 1 of the load cell. It also shows a large offset value (i.e., the output value was about 700 out of 4095 of the resolution of MCP 3208 without weight) owing to magnetic force, which corresponds to around 686 mN preload, which is sufficient for daily tangible materials such as foods and cosmetics. Figure 4 shows the magnetic field distribution of the magnet under the substrate, measured at the surface of the base magnet at 6.25 mm intervals using TM-801 tesla meter (KANETEC, Inc., Nagano, Japan). Although the magnetic force of the magnet under the substrate varies depending on the measurement position, we note that a sufficient preload ranging from 686 to 1323 mN can be exerted at any cells. In other words, this system cannot measure adhesive substances with adhesive force of more than 686 mN at a single pin. While each pin receives the different magnetic force, this value is treated as an offset and the pressure change amount can be measured as a relative value regardless of the initial offset value.

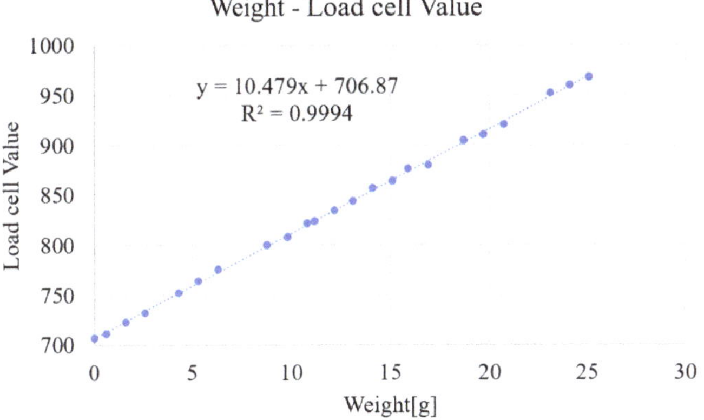

Figure 3. Load cell calibration. We measured load cell value by applying offset pressure by the magnetic force and arbitrary weight.

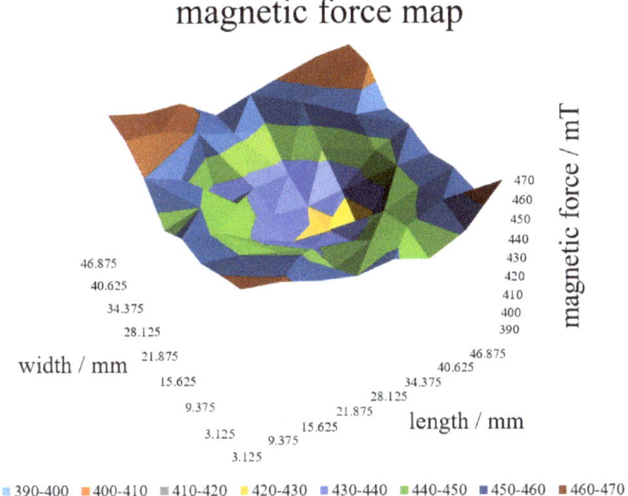

Figure 4. Magnetic force map. The magnetic force was measured at intervals of 6.25 mm along the length and width directions immediately at the surface of the magnet.

2.3. *Preliminary Experiment*

In our stickiness measurement experiments, the adhesive material of interest was applied to a participant's fingertip, who then pressed the fingertip onto the surface of the pin array. When the finger pressure reached 5 N (summed force over all pins), the participant was asked to release the fingertip along the vertical direction. The lifting process was completed in around 1 s. In the study, we used Nattō stirred for 100 times by chopsticks as an adhesive material and baby powder (Johnson & Johnson, Inc. Tokyo, Japan) as a non-adhesive material. Natto is a fermented food in Japan, and when mixed, it becomes sticky.

2.4. *Results and Discussion*

Figures 5 and 6 show the measurement results for Nattō and baby powder, respectively. Although we acquired 2D distributed data, representation by 3D graph was not easy to grasp and we chose to show the 2D view, longitudinal section of the center. As the measurement points are 8 by 12, the horizontal axis of the graphs becomes 1 to 12. The vertical axis represents force (mN), with a positive value meaning tensile force (i.e., negative pressure). In the case of Natto, pressing begins from 0.00 s and finger detachment begins after 2.67 s, totally detached at 4.01 s. In the case of baby powder, pressing begins from 0.00 s, finger detachment begins at 1.67 s, and totally detached at 2.67 s. We asked the participant to release his finger within 1 s, and we visually confirmed that the finger had separated around this time.

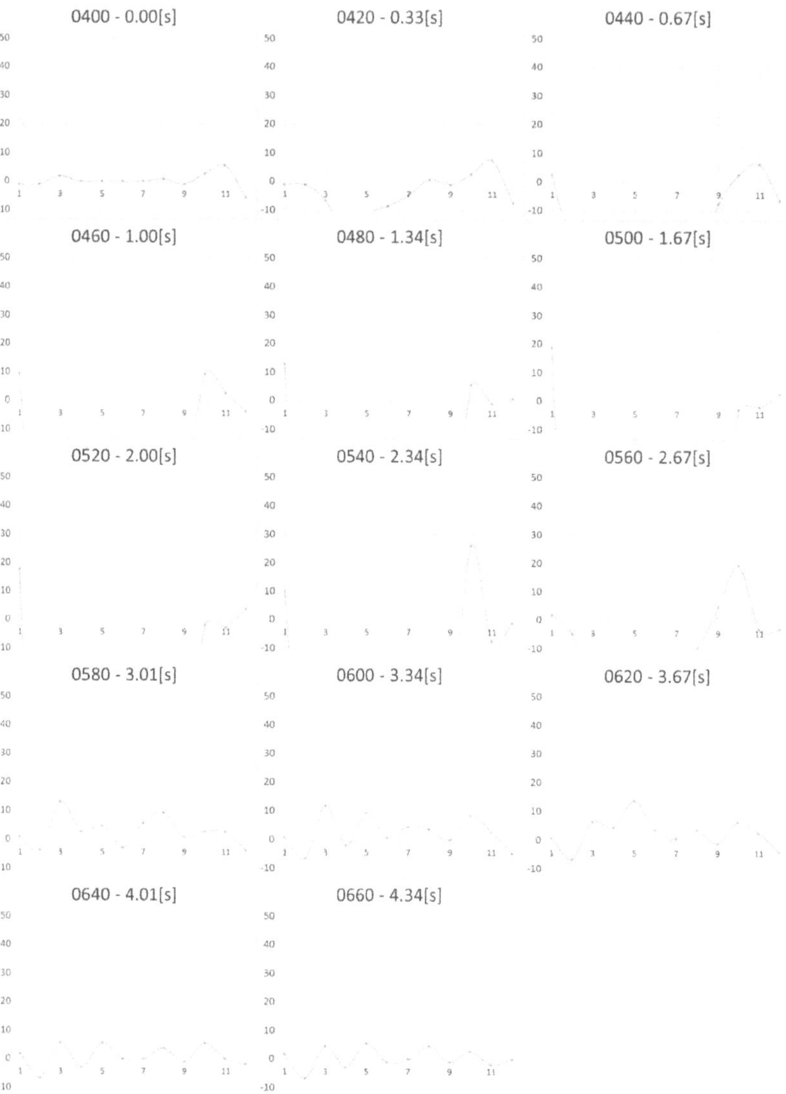

Figure 5. Change in pressure distribution for Nattō (2D view, longitudinal section of the center). The vertical axis represents the force (in milli Newton). The horizontal axis represents the location of the sensing point. The four-digit numbers in each graph show the frame numbers at the time of measurement. Measurements were taken in 0.0165 s/frame (60 fps).

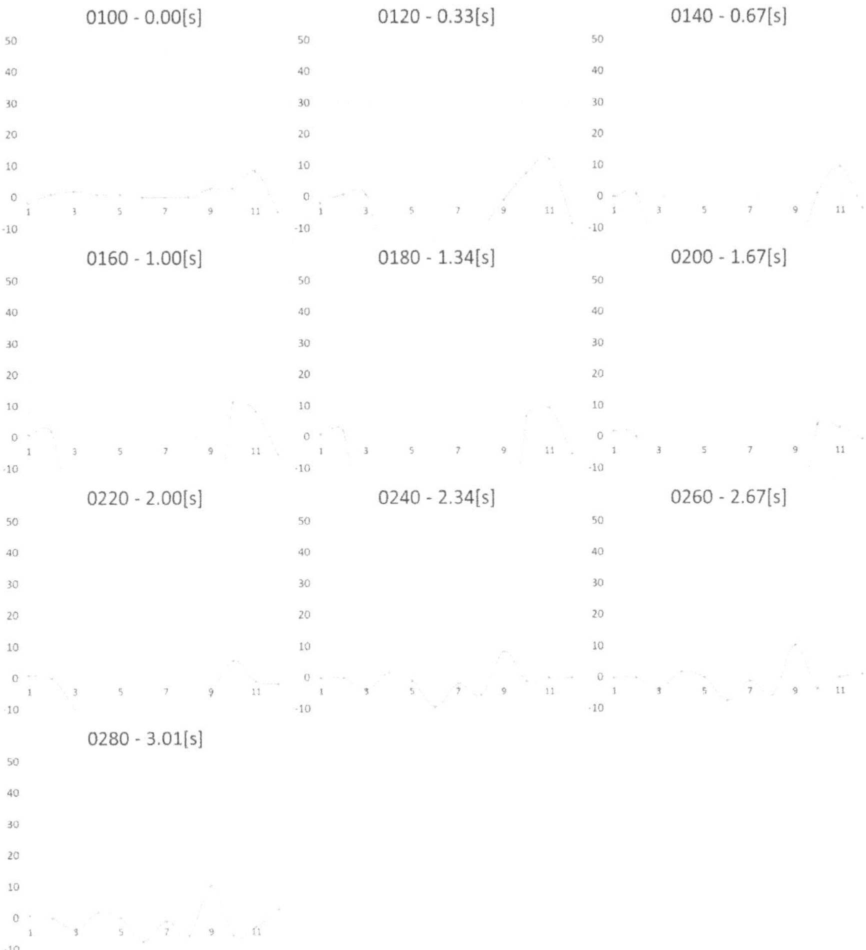

Figure 6. Change in pressure distribution for baby powder (2D view, longitudinal section of the center). The vertical axis represents the force (in milli Newton). The horizontal axis shows the location of the sensing point. The four-digit numbers in each graph show the frame numbers at the time of measurement. Measurements were taken in 0.0165 s/frame (60 fps).

Upon comparing the two figures, we note that peak tensile force is stronger in the case of Nattō. From 2.67 s, it can be observed that the tensile force is generated from the periphery of the surface being pressed and gradually concentrated at the center. In both cases, after the finger was totally detached, we still observe remaining tensile force, which is considered as noise.

One possible explanation of this offset noise is that the pin magnets were aligned along the magnetic lines of the base magnet and the lines were not strictly vertical, which gave rise to the interference between the pin and the pin insertion plate, which may have generated friction and thus noise (Figure 7). This noise can be regarded as a hysteresis of the measuring device.

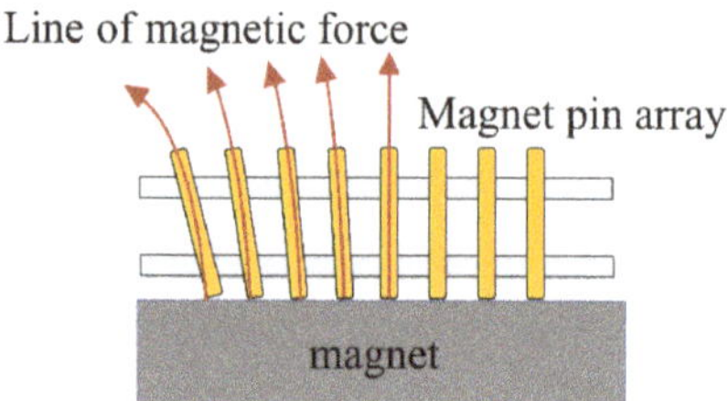

Figure 7. Magnetic lines of force in the setup. The magnetic field lines of the base magnet have a tangential component, generating frictional force between the pin magnet and the pin insertion acrylic plate.

3. Improvement of Measuring System

To solve the problem of the previous prototype—in which the pin magnets were aligned along the base magnet's magnetic lines of force and did not stand vertically, thus potentially causing friction and noise—we next devised a one-row pin-array measuring device. This device is based on the principle that when an infinitely long rectangular magnet is used, the magnetic force line at its center is vertical (Figure 8). Consequently, in the improved device, the pin magnets are installed vertically on the centerline of the rectangular magnet, which reduces the interference with the pin insertion plate (Figure 9). Although only one row can be measured using this configuration when the object in contact with the adhesive substance is semicircular, the distribution of the adhesive force is concentric and measurement at one row is thus sufficient. As we use a human fingertip or artificial human finger here, which can be considered semicircular, we consider a line measurement to be sufficient.

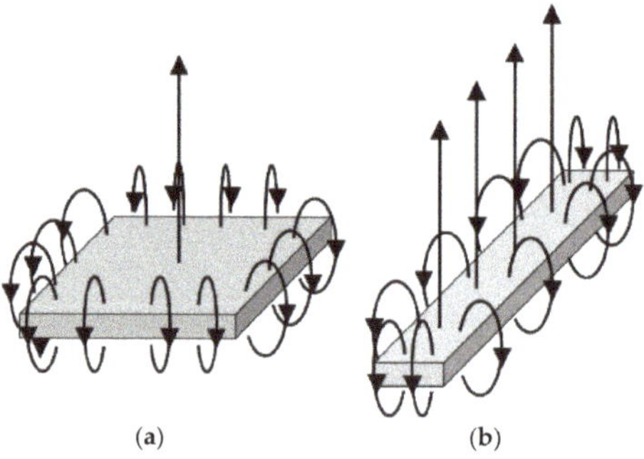

Figure 8. Magnetic field lines of square (**a**) and rectangular magnets (**b**). When an infinitely long rectangular magnet is used, the magnetic force line at its center is vertical.

Figure 9. Pin magnet positioned at center and peripheral regions of rectangular magnet. The photographs indicate that the pin can remain vertical because of vertical magnetic lines.

3.1. Measuring System Using Single-Axis Robot

To apply the offset preload to a load cell, we installed a rectangular magnet such that the center line of the rectangular magnet overlapped with the sensor portion of the single-row load-cell substrate. As a result, the pin magnets were positioned along the vertical lines of the magnetic force. In addition, two rows of pin magnets were added so as to sandwich the measuring pins, although actual measurement was performed only for the pin magnets of the center row. These additional rows of pins push the central pins from both sides with magnetic repulsive force, making the central pins perpendicular to the base surface and thus minimizing friction between the pin insertion plate and the pins. The actual device is shown in Figure 10. The device consists of an acrylic pin insertion plate, a 3 × 16 magnet pin array, two 100 mm × 20 mm × 7 mm magnets, and a 1 × 16 load cell substrate. Two base magnets were stacked to strengthen the magnetic field, which was about 410 mT at the center. The load cell and its surrounding circuit components are the same as in the previous setup, while the load cell substrate was redesigned to achieve 1 × 16 load cell configuration. Each pin of the magnet pin array was aligned to correspond to one load cell, and a preload was applied to the load cells.

The load cells were arranged at intervals of 2.54 mm, and the magnet pin array was arranged in the same manner. Each pin magnet was 2 mm in diameter and 10 mm in height, and the magnetic force was 275 mT, corresponding to a preload of about 980 mN.

For contact with the adhesive substance, we used a hemispherical artificial human skin gel (with a diameter of 5 cm, manufactured by BEAULAX Corporation, Saitama, Japan) with elasticity equivalent to that of human skin as the contactor along with a single-axis robot (T4L manufactured by YAMAHA, Shizuoka, Japan) to depress the artificial "finger".

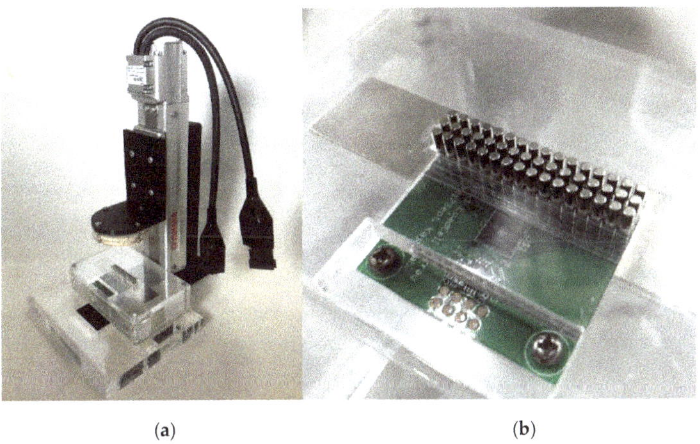

(a) (b)

Figure 10. Single-column pin-array measuring device (**a**) overall view, (**b**) magnified view of the measurement part.

3.2. Experiment 1

Adhesive substances were uniformly applied to the upper part of the pin array beforehand, and the contactor was pressed against the upper surface of the pin array placed on the load cell. We placed 1 mm thick adhesive materials on each pin, with the procedure depicted in Figure 11. The temporal change in the pressure distribution was measured when pulling apart the contactor from the surface. The single-axis robot was used for pressing and separating the "finger", and the pushing distance of the contactor was set to 2 mm vertically downward from the state in contact with the pin array, while the pulling-off distance was set to 2 mm vertically upward from the state in contact with the pin array. The moving speed of the contactor was 1 mm/s. Honey, toothpaste, shaving gel, and shampoo were prepared as adhesive substances. These adhesive substances are familiar to us, and we can find difference by touching them. If this system can measure the difference in adhesive force of these substances, there is a possibility that the system can grasp the difference in the adhesive feeling felt by human skin. The sticky substance attached to the contactor and pin arrays was removed each time before the next trial.

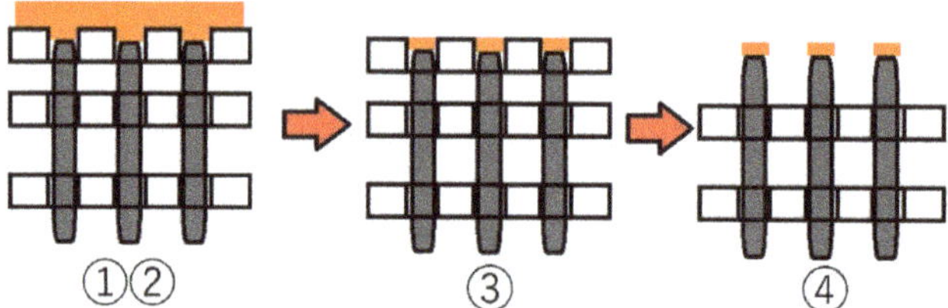

Figure 11. Procedure of pasting adhesive material on each pin. ① Place the insertion plate over the pin magnet, ② coat with an adhesive sample, ③ remove excess adhesive samples, and ④ remove the insertion plate.

3.3. Results and Discussion

The measurement results in the case of honey are shown in Figure 12. The horizontal axis represents the pin location, whereas the vertical axis represents the force. A positive value indicates a tensile force (i.e., negative pressure). From the figure, we note that pushing starts at 0 s, whereas pulling begins at around 3.00 s. A tensile (i.e., adhesive) force is observed at the edge of the contact surface. It can also be confirmed that the adhesive force transfers to the center as the contact surface area changes with the motion of the contactor. Overall, the noise was reduced from the previous prototype, and we now can clearly observe tensile force behavior.

On the other hand, we still have some issues. When comparing the pressure distribution after measurement (5.00 s) and before measurement (0 s), we observed a residual force in the positive direction after the measurement. The reason for this hysteresis phenomenon is unclear (we have confirmed that the load cell itself does not have observable hysteresis), but we presume that friction between the contact pin and the plate still existed. Further, a tensile force can be observed to the leftmost section of the graph in the interval from 1.33 s in Figure 12, which should not have contacted the contactor and should be regarded as noise. As the rightmost and leftmost sensing points are not surrounded by other pin magnets, they experience a magnetic force from the neighboring pin magnets to generate a repulsive force in the left and right directions and interfere with the pin insertion plate.

Figures 13–15 show the measurement results for toothpaste, shaving gel, and shampoo, respectively. As in the case of honey, we were able to measure the change in adhesion. When the results were compared for each adhesive sample, it was found that in the case of toothpaste, the adhesive force was the strongest and that the viscosity was high and the duration of the adhesive force was long, together with the fact that the adhesive force remained up to 5.0 s. The shaving gel and shampoo had weak adhesion.

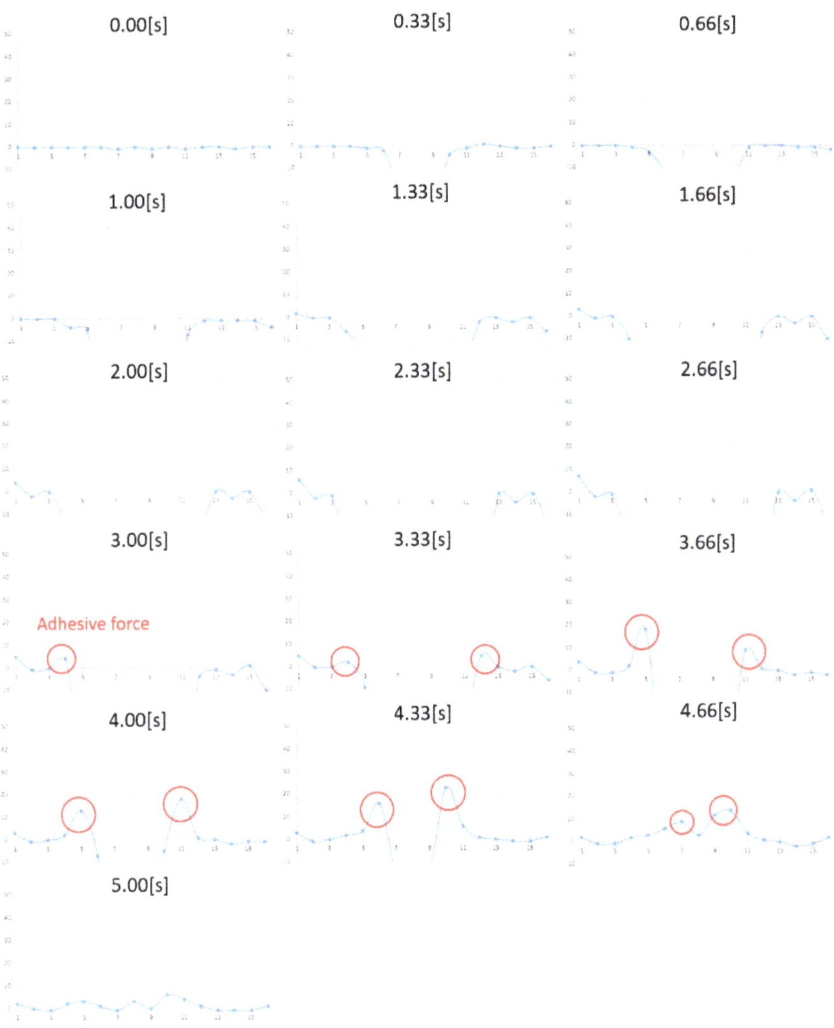

Figure 12. Change in pressure distribution for honey. The vertical axis represents the force (in milli newton). The horizontal plane represents the location of the sensing point.

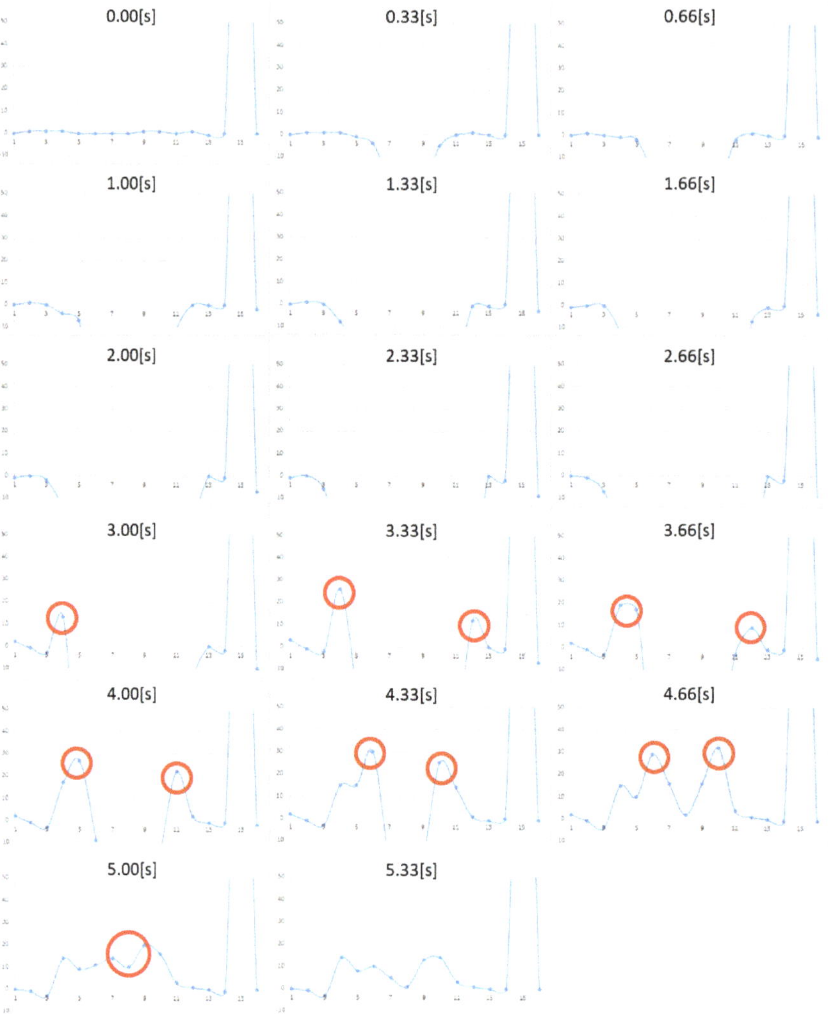

Figure 13. Change in pressure distribution for toothpaste. The vertical axis represents the force (in milli newton). The horizontal plane represents the location of the sensing point.

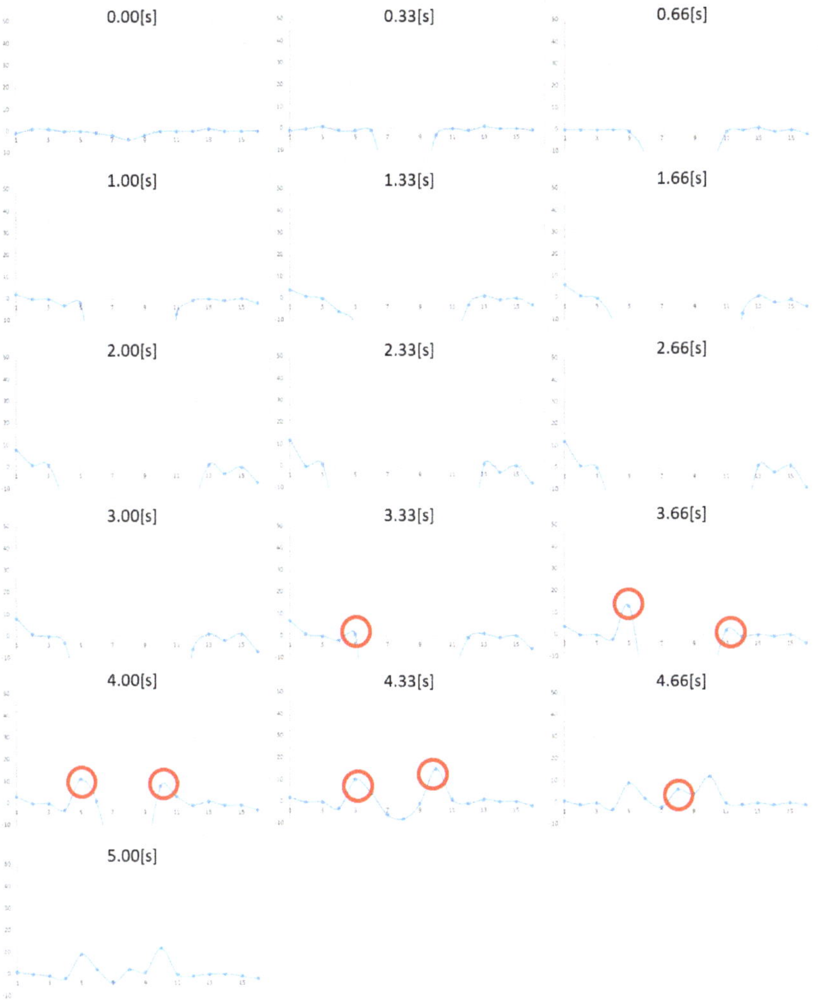

Figure 14. Change in pressure distribution for shaving gel. The vertical axis represents the force (in milli newton). The horizontal plane represents the location of the sensing point.

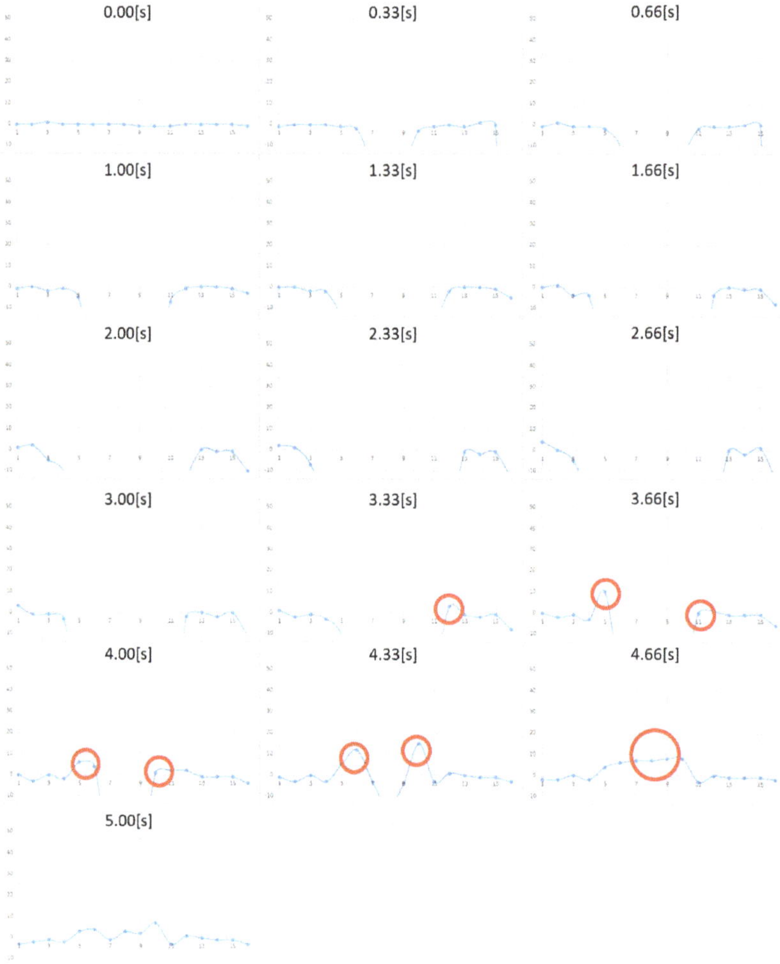

Figure 15. Change in pressure distribution for shampoo. The vertical axis represents the force (in milli newton). The horizontal plane represents the location of the sensing point.

3.4. Experiment 2

Subsequently, in order to verify whether the difference in adhesive force could be measured by this system, we carried out measurements using a standard viscosity liquid. In this experiment, kinematic viscosities (Centi-Stokes Visco Liquid, ASONE, Inc., Tokyo, Japan) of 500, 1000, 3000, 5000, and 10,000 (cSt) were measured. The moving speed and force of the contact object in the measurement were set to values identical to the corresponding ones in experiment 1. The peak adhesive force during one measurement was acquired 10 times consecutively for each sample, and the average value was used as the measurement data of each sample.

3.5. Results and Discussion

Figure 16 shows the peak value of the adhesive force for each sample as acquired from the measured data. As can be observed from the figure, there is an obvious correlation between the peak values of the adhesive force and kinematic viscosity, and the difference in the adhesive force can be suitably measured by means of our measuring device. The peak value of the adhesive force becomes constant when the kinematic viscosity exceeds 5000 cSt.

The purpose of fabricating the second prototype was to solve the issues of the first prototype, specifically the large observed noise possibly owing to friction between the pins and the plate caused by inclined magnetic lines. Our results suggest that this noise is reduced with the new setup, and the tensile force distribution can be clearly observed. This in turn implies that our speculation of the cause of the noise was correct.

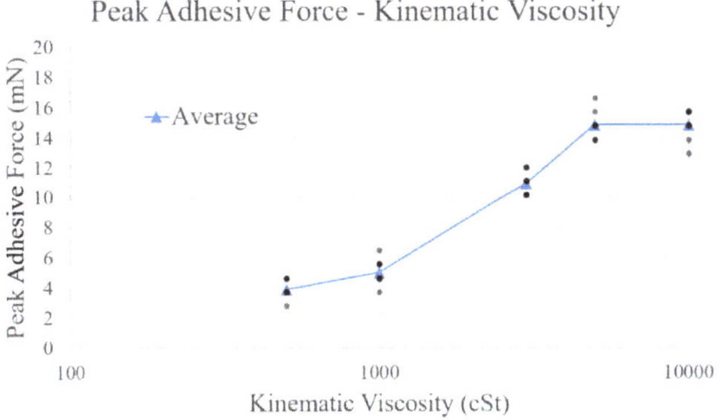

Figure 16. Plot of peak values of adhesive force (mN) and kinematic viscosity (cSt). The blue line indicates the average peak value, whereas the gray points indicate the row data (for standard viscosity liquids of 500, 1000, 3000, 5000, and 10,000 cSt).

4. Conclusions

In our study, we developed a measuring device for quantifying stickiness. A pressure distribution sensor was used to observe the temporal change in the pressure distribution upon applying finger pressure to a given adhesive material. Here, we note that a typical pressure distribution sensor can measure a pressing force, but not a tensile force, and thus we proposed and implemented a method of measuring the tensile force by applying an offset pressure in advance to the sensor and measuring the resulting difference upon finger pressure application.

Subsequently, we developed a pressure distribution sensor board with built-in load cells and independent sensing points for highly accurate measurements and implemented a preload application method using a magnetic force. We next compared measurements acquired using a fingertip coated with Nattō as an adhesive and a fingertip coated with baby powder. The results confirmed that the tensile force generated by the adhesive substance was initially at the edge of the contact part, but moved with the change in contact pressure, and eventually converged to the center.

However, owing to the inclination of the magnetic lines of force, there was interference between the pin magnet and the pin insertion plate. Therefore, we proposed a linear arrangement to simplify the system and provide a stable vertical preload. We subsequently measured several kinds of adhesive substances using a single-axis robot that pressed an artificial finger onto the device coated with the sticky substance of interest. The results indicated that the new setup was able to measure the adhesive force distribution more accurately. We believe that our device can be used to suitably quantify the stickiness of adhesive substances.

The current obvious limitation is that we did not fully eliminate noise. Our method also has some innate drawbacks, such as the fact that it cannot measure adhesive force exceeding the offset pressure, and excessively low viscosity substances such as water cannot be treated because they fall from the magnet pins. Still, we believe that the measurement of adhesive force distribution should contribute to the study of stickiness, and comparing our measured data with a human's subjective tactile feeling is our next step.

Author Contributions: Conceptualization, T.K. and H.K.; methodology, T.K. and H.K.; software, T.K. and A.T.; validation, K.M., N.S., and N.A.; formal analysis, T.K., A.T., and V.Y.; investigation, T.K.; resources, H.K., K.M., N.S., and N.A.; data curation, T.K.; writing—original draft preparation, T.K.; writing—review and editing, T.K. and H.K.; visualization, T.K.; supervision, H.K.; project administration, H.K.; funding acquisition, H.K.

Funding: This research was funded by JSPS Grant-in-Aid for Scientific Research, grant number 15H05923 (New academic field research "Multiple quality sensing").

Conflicts of Interest: The authors declare no conflict of interest.

References

1. Hayward, V.; Levesque, V. Experimental evidence of lateral skin strain during tactile exploration. In Proceedings of the Eurohaptics, Dublin, Ireland, 6–9 July 2003.
2. Bicchi, A.; Scilingo, E.P.; and De Rossi, D. Haptic discrimination of softness in teleoperation: the role of the contact area spread rate. *IEEE Trans. Rob. Autom.* **2000**, *16*, 496–504. [CrossRef]
3. Kimura, F.; Yamamoto, A.; Higuchi, T. Development of a contact width sensor for tactile tele-presentation of softness. In Proceedings of the 18th IEEE International Symposium on Robot and Human Interactive Communication, Toyama, Japan, 27 September–2 October 2009; pp. 34–39.
4. Levesque, V.; Jerome, P.; Hayward, V. Braille Display by Lateral Skin Deformation with the STReSS2 Tactile Transducer. In Proceedings of the Second Joint EuroHaptics Conference and Symposium on Haptic Interfaces for Virtual Environment and Teleoperator Systems (WHC'07), sukaba, Japan, 22–24 March 2007; pp. 115–120.
5. Schorr, S.B.; Okamura, A.M. Three-Dimensional Skin Deformation as Force Substitution: Wearable Device Design and Performance During Haptic Exploration of Virtual Environments. *IEEE Trans. Haptic* **2017**, *10*, 418–430. [CrossRef] [PubMed]
6. Fujita, K.; Ikeda, Y. Remote haptic sharing of elastic soft objects. Proceedings of The First Joint Eurohaptics Conference and Symposium on Haptic Interface for Virtual Environment and Teleoperator Systems, Pisa, Italy, 18–20 March 2005.
7. Endo, T.; Kusakabe, A.; Kazuma, Y.; Kawasaki, H. Haptic Interface for Displaying Softness at Multiple Fingers: Combining a Side-Faced-Type Multifingered Haptic Interface Robot and Improved Softness-Display Devices. *IEEE/ASME Trans. Mechatron.* **2016**, *21*, 2343–2351. [CrossRef]
8. Bau, O.; Poupyrev, I.; Isrer, A.; Harrison, C. TeslaTouch: Electrovibration for Touch Surfaces. In Proceedings of the 23nd annual ACM symposium on User interface software and technology, New York, NY, USA, 3–6 October 2010; pp. 283–292.
9. Chen, X.; Shao, F.; Barnes, C.; Childs, T.; Henson, B. Exploring Relationships between Touch Perception and Surface Physical Properties. *Int. J. Des.* **2009**, *3*, 67–76.
10. Tiest, W.M.B.; Kosters, N.D.; Kappers, A.M.L.; Daanen, H.A.M. Haptic perception of wetness. *Acta Psychol* **2012**, *141*, 159–163. [CrossRef] [PubMed]
11. Filingeri, D.; Fournet, D.; Hodder, S.; Havenith, G. Why wet feels wet? A neurophysiological model of human cutaneous wetness sensitivity. *J. Neurophysiol.* **2014**, *112*, 1457–1469. [CrossRef] [PubMed]
12. Fisher, T.H. What we touch, touches us: Materials, affects, and affordances. *Des. Issue.* **2004**, *20*, 20–31. [CrossRef]
13. Liu, H.; Bhushan, B. Adhesion and friction studies of microelectromechanical systems/nanoelectromechanical systems materials using a novel microtriboapparatus. *J. Vac. Sci. Technol. A* **2003**, *21*, 1528–1538. [CrossRef]
14. Yamaoka, M.; Yamamoto, A.; Higuchi, T. Basic Analysis of Stickiness Sensation for Tactile Displays. In *Computer Science*; Springer: Berlin, Heidelberg, Germany, 2008; pp. 427–436.
15. Angioloni, A.; Collar, C. Bread crumb quality assessment: a plural physical approach. *Eur. Food Res. Technol.* **2009**, *229*, 21–30. [CrossRef]
16. Liu, Z.; Scanlon, M.G. Modelling Indentation of Bread Crumb by Finite Element Analysis. *Biosystems Eng.* **2003**, *85*, 477–484. [CrossRef]
17. Dan, H.; Watanabe, H.; Dan, I.; Kohyama, K. Effects of textural changes in cooked apples on the human bite, and instrumental tests. *J. Texture Stud.* **2003**, *34*, 499–514. [CrossRef]

18. Kameoka, T.; Takahashi, A.; Yem, V.; Hiroyuki, K. Measurement of Stickiness with Pressure Distribution Sensor. In *Haptic Interaction*; Springer: Singapore, 2016; pp. 315–319.
19. Kameoka, T.; Takahashi, A.; Vibol, Y.; Kajimoto, H. Quantification of stickiness using a pressure distribution sensor. In Proceedings of the 2017 IEEE World Haptics Conference (WHC), Munich, Germany, 6–9 June 2017; pp. 617–622.
20. Sidney, W. *Intensive and Extensive Aspects of Tactile Sensitivity as a Function of Body Part, Sex and Laterality*; Thomas International Publishing: New York, NY, USA, 1968; pp. 195–222.

Article

Biomimetic Tactile Sensors with Bilayer Fingerprint Ridges Demonstrating Texture Recognition

Eunsuk Choi [1], Onejae Sul [2], Jusin Lee [1], Hojun Seo [1], Sunjin Kim [1], Seongoh Yeom [1], Gunwoo Ryu [1], Heewon Yang [1], Yoonsoo Shin [1] and Seung-Beck Lee [1,2,*]

[1] Department of Electronic Engineering, Hanyang University, 222 Wangsimni-ro, Seongdong-gu, Seoul 04763, Korea; silver77@hanyang.ac.kr (E.C.); jusin19@hanyang.ac.kr (J.L.); masiks@hanyang.ac.kr (H.S.); akangel0307@gmail.com (S.K.); yso526@hanyang.ac.kr (S.Y.); rgwrgw00@gmail.com (G.R.); heewon0820@hanyang.ac.kr (H.Y.); skstls715@hanyang.ac.kr (Y.S.)

[2] Institute of Nano Science and Technology, Hanyang University, 222 Wangsimni-ro, Seongdong-gu, Seoul 04763, Korea; ojsul@hanyang.ac.kr

* Correspondence: sbl22@hanyang.ac.kr; Tel.: +82-2-2220-1676; Fax: +82-2-2294-1676

Received: 2 September 2019; Accepted: 23 September 2019; Published: 25 September 2019

Abstract: In this article, we report on a biomimetic tactile sensor that has a surface kinetic interface (SKIN) that imitates human epidermal fingerprint ridges and the epidermis. The SKIN is composed of a bilayer polymer structure with different elastic moduli. We improved the tactile sensitivity of the SKIN by using a hard epidermal fingerprint ridge and a soft epidermal board. We also evaluated the effectiveness of the SKIN layer in shear transfer characteristics while varying the elasticity and geometrical factors of the epidermal fingerprint ridges and the epidermal board. The biomimetic tactile sensor with the SKIN layer showed a detection capability for surface structures under 100 μm with only 20-μm height differences. Our sensor could distinguish various textures that can be easily accessed in everyday life, demonstrating that the sensor may be used for texture recognition in future artificial and robotic fingers.

Keywords: biomimetic; tactile sensor; fingerprint ridge; piezoelectric sensor; texture discrimination

---

1. Introduction

The human ability of tactile sensing using the finger plays an essential role in object manipulation and our interaction with the external environment. When making contact with an object, we can recognize its texture using tactile senses made through mechanoreceptors distributed throughout the skin [1]. Therefore, research into tactile sensing attempts to mimic the mechanoreceptors in the human skin, their pressure sensing ability, and their distribution in the skin. Recently, many groups have reported on pressure sensors that demonstrate basic artificial tactile feeling and texture pattern recognition for prosthesis and robotics applications. They have placed pressure sensors at the digits of robotic arms to grip and manipulate objects and have arranged pressure sensors in an array form to detect the shape of an object [2–5]. Electronic skin instrumented with pressure sensors for prosthesis application has also been reported [6–8].

Many attempts at improving tactile sensing ability have focused on enhancing pressure sensing sensitivity in the belief that higher pressure sensitivity would lead to enhanced sensibility of the contacting object. To enhance sensitivity of the pressure sensors, a capacitor-type pressure sensor was demonstrated, which was able to detect only a few milligrams of tactile pressures [9–15]. However, the high pressure sensitivity was achievable only at low spatial resolution, limiting the fine-texture detecting capability. A closer inspection of the human skin showed that the sensitivity and distribution of the mechanoreceptors are not the only criteria for tactile sensing.

When a human finger slides over an object's surface, the fingerprint ridge structures interact with the object's surface structure and generate vibrational signals, which are then transmitted to the mechanoreceptors lying under the epidermis. The generated action potential pulses in the mechanoreceptors travel to the brain, where they are interpreted as shear interaction and are used for recognizing the "fine texture" of the object [16]. To detect fine surface textures, ridge structures mimicking human fingerprint ridges have been applied to tactile sensors [17–21]. The biomimetic ridge structures showed the ability to detect a few hundred micrometers of scale surface structures [18], and it was found that biomimetic ridge structures are an essential element for material and textural identification. Despite there being many studies on tactile sensors with biomimetic ridge structures, only a few have reported studies on the ridge structure itself, e.g., elasticity, dimension, for the improvement of texture recognition.

In this study, we developed a biomimetic tactile sensor with a surface kinetic interface (SKIN) that imitates two elements of the human skin: epidermal fingerprint ridges and the epidermis. Our tactile sensor is composed of a SKIN (biomimetic epidermal fingerprint ridges and supporting epidermal board) layer and a polyvinylidene difluoride (PVDF) piezoelectric polymer layer that acts as a vibration sensor, which functions similarly to a fast-adapting mechanoreceptor. We evaluated the effectiveness of the SKIN layer in transferring surface shear forces into vertical vibrations while varying the elasticity and geometrical factors of the epidermal fingerprint ridges and epidermal board. We found that the tactile sensitivity was increased when using a softer epidermal board and a harder epidermal fingerprint ridge. The biomimetic tactile sensor with SKIN showed detectability for sub-100-μm surface structures and 20-μm height differences. Our sensor produced signals that made it possible to discern different types of paper, leather, and fabrics. It was also possible to distinguish a real human finger from its replica. The developed biomimetic tactile sensor may be applied to future artificial fingers for texture detection and recognition.

2. Surface Kinetic Interface (SKIN)

An illustration of human skin on a fingertip can be seen in Figure 1a. From a tactile sensing perspective, the human skin is composed of two parts, the epidermis and the dermis. The epidermis comes in contact with the outside world and transfers kinetic interactions to the dermis. In the dermis, various mechanoreceptors convert the mechanical stimulus into action potential signals. The mechanoreceptor illustrated in Figure 1a is a Meissner's corpuscle, which generates an action potential responding to changes in the surface strain [22]. In the finger epidermis, the top layer forms epidermal fingerprint ridges (ERs) that interact directly with the contact material. Underneath it, there are layers of cells that we call the epidermal board that collectively act as a force transfer layer [23]. We designed our tactile sensor skin to functionally mimic the outer skin of the human fingertip: the SKIN layer that consists of ERs on top of a supporting epidermal board. The PVDF piezoelectric thin-film layer acts as an array of vibration sensors mimicking fast-adapting mechanoreceptors (see Figure 1b). The way shear forces are detected in our tactile sensor is quite similar to the human fingertip. When an object horizontally touches the SKIN surface, a lateral shear force is applied to the ERs. Torque is then generated at the contact point between the ERs, with the epidermal board acting as an axis of rotation. Therefore, the contact shear force is transformed into torque and then transferred as vertical pressure to the PVDF layer. The PVDF layer is a piezoelectric material that generates a voltage signal when applied pressure deforms its polarized molecular structures. The magnitude of the generated voltage signal depends on the magnitude of the lateral shear force; as higher shear force induces higher vertical pressure on the PVDF. In addition, the frequency of the voltage signal depends on the frequency of contact with the ERs. This depends on the relative speed of the scanning object and the period of surface protrusions on the object's surface that makes contact with the ERs. Therefore, the magnitude and frequency of the generated output voltage depend mostly on the surface roughness and hardness. This will be further elaborated in Section 4. Although our tactile sensor is biomimetic in the SKIN structure, there are differences from human skin in that unlike each mechanoreceptor individually detecting the input stimulus, the PVDF layer acts as a continuous distribution of sensors, with the detected output signal being a parallel integration of all the voltage signals generated over the whole surface of the sensor at that particular time interval. In mapping the surface topography of a contacting material, this would be disadvantageous, since any information generated through the spatial distribution of contact pressure points within the sensor area would not be distinguishable. However, it will be shown that for detecting the texture of the contacting surface, the parallel integration of the voltage signals becomes advantages. This sensing mechanism is similar to vibrotactile sensation in the duplex theory of human tactile perception using fast-adapting mechanoreceptors [24].

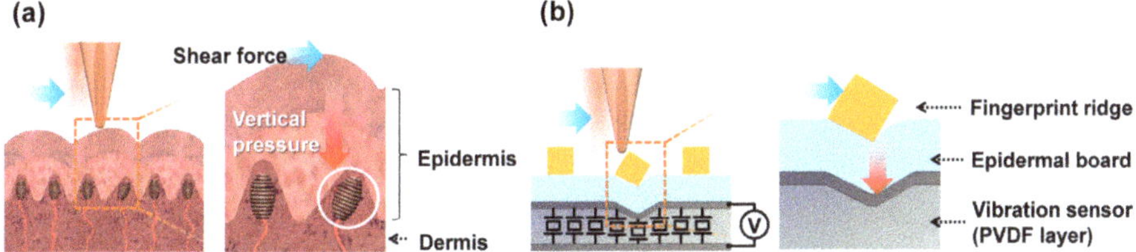

Figure 1. Schematic illustration and shear detecting mechanism of (**a**) the human fingertip and (**b**) the biomimetic tactile sensor.

3. Fabrication

The Young's modulus of human fingerprint ridges is known to be about three times higher than that of the epidermal board, which is believed to increase the durability of the outer skin layer [25]. To mimic this configuration of the human epidermis, we fabricated harder ERs using SU-8 (Microchem, Westborough, MA, USA), which had a Young's modulus of ~3 GPa, and a soft epidermal board (using polydimethylsiloxane (PDMS)), which had a Young's modulus of ~750 kPa. For efficient force transfer, ERs should not absorb the contact force through elastic deformation, while the epidermal board should be flexible enough to localize the deformation and allow the ER to tilt with the applied contact shear force. To test the effectiveness of this configuration, we fabricated four types of SKINs with differing mechanical configurations: hard or soft ERs and hard or soft epidermal boards. We used SU-8 and polyethylene terephthalate (PET) as the harder material ($E \approx 3$ GPa) and PDMS as the softer material ($E \approx 750$ kPa) (see Figure 2a). The geometry of all SKINs was identical, with the thickness of the epidermal board at 50 μm, the height of the ER at 50 μm, and the width of the ER plateau and valley both at 100 μm, giving the fingerprint pattern a 200-μm period, which is roughly half of the period of human fingerprints (0.3–0.5 mm) [26].

Figure 2b shows the fabrication processes of the different SKINs. For SKIN #1, an epidermal board was fabricated by spin-coating 50-μm-thick SU-8 on an oxidized silicon substrate. Then the SU-8 2075 was coated again at a 50-μm thickness, and ERs were patterned by using optical lithography. Then the fabricated SKIN #1 was lifted off through HF etching of the oxide underneath. For SKIN #2, a 50-μm-thick SU-8 ER master was patterned using optical lithography. A PET film was attached to the ER master and was dipped into the PDMS. The viscosity of the PDMS was low enough to fill the vacant spaces between the SU-8 master and the PET film. After the PDMS was cured, the fabricated SKIN #2 was peeled off from the master. The PET film for SKIN #2 was etched by CF_4 physical plasma to form nanobrush structures, which promoted adhesion with PDMS by increasing the contact surface area [27,28]. For SKIN #3, a 50-μm-thick SU-8 layer and 50-μm-thick PDMS were coated on the oxidized silicon substrate. This SU-8 layer was helpful for the separation of the PDMS layer from the substrate after completion of SKIN #3. Then the surface of the PDMS was treated by oxygen plasma for better adhesion with a 2-μm-thick SU-8 layer coated on PDMS. This SU-8 thin layer works as the adhesive layer between the PDMS and the following SU-8 ERs. A 50-μm-thick SU-8 ER was patterned by using optical lithography. Finally, the fabricated SKIN #3 was peeled off from the bottom SU-8 layer on the substrate. For SKIN #4, a 50-μm-thick SU-8 ER master was patterned, and 50-μm-thick PDMS was coated on the ER master. After the PDMS was cured, SKIN #4 was peeled off.

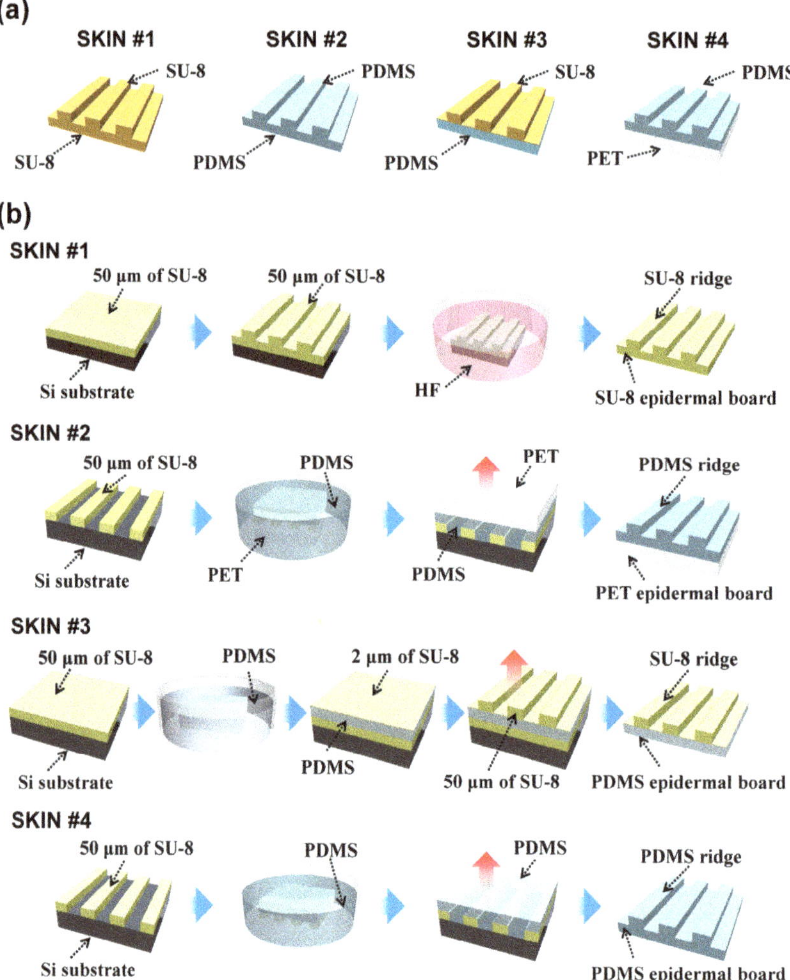

Figure 2. (**a**) Four types of surface kinetic interfaces (SKINs) with differing mechanical configurations and (**b**) their fabrication processes.

Figure 3a shows an SEM image of the fabricated SKIN #3. The fabricated SKIN was attached to a 28-μm-thick PVDF sensor (DT1-028K, Measurement Specialties, Hampton, VA, USA) to complete the flexible tactile sensor device structure (Figure 3b). The PVDF sensor was 12 mm wide and 30 mm long and had a piezoelectric voltage constant (g_{33}) of −0.33 Vm/N. Figure 4 shows the measurement set-up, which had a 2-axis (x–z) motorized stage with 2 μm resolution. The scanning speed was controlled by the motorized *x* axis, and the contact depth was controlled by a motorized *z* axis. A PET tip with 5 mm of width and 125 μm in thickness was used for applying shear force to the ER (see inset). The PET tip was scanned across the SKIN surface without any vertical pressure, and the response voltage signal of the PVDF sensor was monitored using an oscilloscope. The tactile signal was obtained by subtracting the reference noise (which was obtained when the motorized stage was scanned above a sample without touching) from the measured PVDF signal in the frequency domain [29].

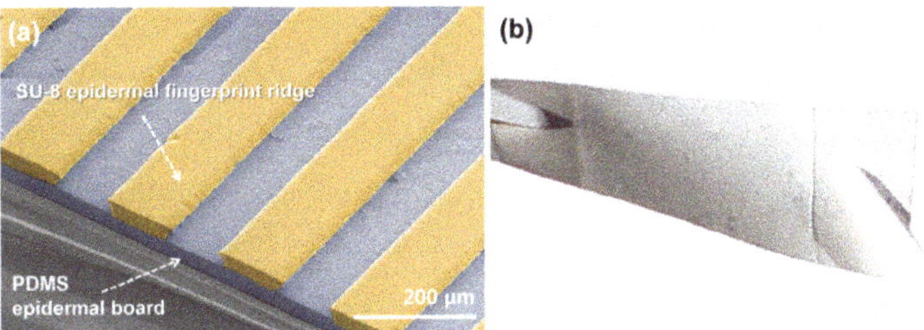

Figure 3. Images of a fabricated biomimetic tactile sensor with SKIN: (**a**) a false-color SEM image of the fabricated SKIN #3, which was composed of (yellow) SU-8 epidermal fingerprint ridges with a 200-μm-period and a 50-μm height, and a (blue) polydimethylsiloxane (PDMS) epidermal board with a 50-μm thickness; (**b**) an optical image of the fabricated flexible biomimetic tactile sensor with SKIN.

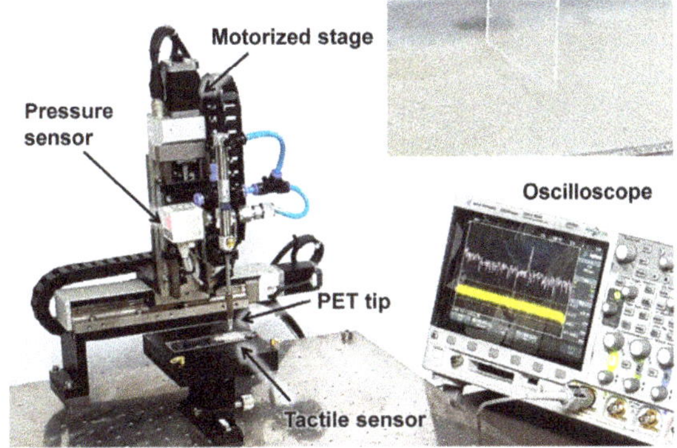

Figure 4. Optical image of the measurement set-up, where the inset shows an optical image of a SKIN and a hovering polyethylene terephthalate (PET) tip.

4. Tactile Sensitivity Dependent on SKIN Configuration

The shear-induced torque applied to the ER structure acts to amplify the contact information [17,18]. To test the effect of having an ER on the sensor surface, the PET tip was scanned at 2.5 cm/s on the tactile sensor with SKIN #1 and the tactile sensor with only an SU-8 epidermal board without an ER. In the case of the tactile sensor with an ER, a significant fluctuation was observed in the sensor output signal after 0.2 s (see Figure 5a). Figure 5b shows the sensor signal between 0.3 and 0.4 s. It was observed that the main period of signal fluctuation was about 8 ms. However, in the case of the tactile sensor without ER, a periodic signal was not observed except for a 15-mV fluctuation induced by power noise (see Figure 5c). For more detailed analysis, we converted the time-dependent PVDF output signal to the frequency domain using fast Fourier transform (FFT). Figure 5d shows the denoised FFT results of Figure 5b,c. The tactile sensor without ERs showed frequency spectra in the low frequency range. This feature was induced by the stick-slip of the PET tip on the epidermal board. However, it was observed that there was a noticeable peak around 125 ± 5 Hz from the tactile sensor with an ER. Judging from the fact that $f = v/p$, where f is the signal frequency, v is the scanning speed, and p is the structural period, the distinct peak of about 125 Hz was induced by periodical interaction between the PET tip and the ER. This result shows that the ERs generate periodic contact information at a specific frequency. We defined this peak frequency induced by the period of ERs and the scanning speed as f_{ER}.

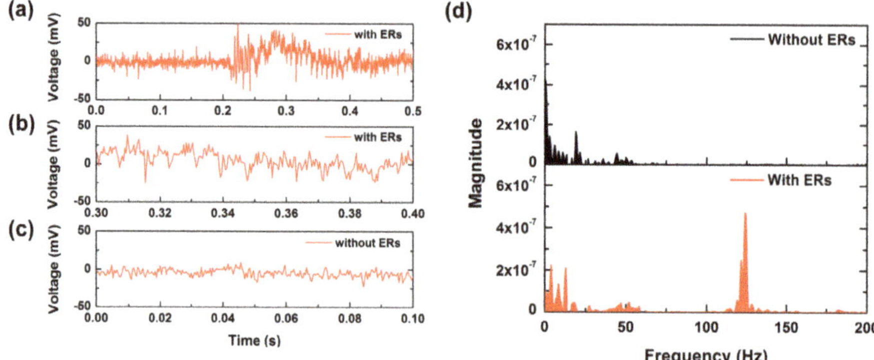

Figure 5. Measurement results of tactile sensor with or without epidermal fingerprint ridges (ERs) when a PET tip was scanned at 2.5 cm/s on the tactile sensor: (**a**) the polyvinylidene difluoride (PVDF) output signal of a tactile sensor with ridge structure; (**b**) the region of 0.3–0.4 s in (a); (**c**) the PVDF output signal of a tactile sensor without ridge structure; (**d**) the fast Fourier transform (FFT) results of (b,c).

To test which SKIN layer composition had the most efficient transfer of surface interactions to the PVDF sensor, we performed the PET tip sliding measurement over the four different SKIN types and compared their FFT results. It can be seen in Figure 6 that all of the SKINs showed a peak in the FFT spectrum around f_{ER} = 125 Hz, which corresponded with the 2.5 cm/s scanning speed of the PET tip. The slight variation in the peak frequency may have been due to swelling in the polymer layers that was created during the different fabrication processes, causing expansion in the epidermal boards. In comparing the magnitudes of f_{ER}, we found that SKIN #3 gave the highest magnitude of −60.50 dB, followed by SKIN #1 at −64.90 dB: both had SU-8 ERs. The harder SU-8 would have less elastic deformation than PDMS would, absorbing less of the lateral shear force applied by the sliding PET tip. Hence, as the PET tip slid over the SU-8 ERs, its interaction was more pronounced than with the SKINs with PDMS ridge structures, resulting in the signal magnitude of f_{ER} being 6.85 times higher for the sensor in SKIN #3 than in SKIN #2 (−68.86 dB), which both had PDMS epidermal boards. Comparing SKIN #3 and SKIN #1, the signal magnitude of f_{ER} was 2.75 times lower for the sensor with SKIN #1, which had the harder SU-8 epidermal board. This is understandable if one considers that a softer epidermal board would allow the ERs to tilt more with the applied shear force and create higher local deformation, while the harder SU-8 would distribute the vertical force over a wider area with lower local deformation, which would reduce the sensor signal. Therefore, it is reasonable that SKIN #4 (with soft PDMS ERs and a hard PET epidermal board) showed the lowest magnitude of f_{ER}. The configuration of the material mechanical properties of SKIN #3 best mimicked that of the human epidermis (with harder fingerprint ridges on top of a softer epidermal board), which gave it the highest amplifying ability. From these results, we found that the human epidermal structure is designed not only to increase ductility but also to efficiently transfer the input stimulus on the skin to the mechanoreceptors.

The heights of the ER structures could also affect the amount of shear force transferred to the PVDF sensor, since the generated torque would depend on the length of the displacement vector. To investigate the dependence of the sensor's sensitivity on the height of the ER structure, we performed PET tip scanning measurements on the tactile sensor using SKIN #3, with ridges of 25 μm, 50 μm, 75 μm, and 100 μm in height. The f_{ER} of a pronounced peak appeared in the FFT results of all four tactile sensors at 125 Hz, corresponding to the scanning speed of 2.5 cm/s. Figure 7a shows the magnitude of f_{ER} of all four tactile sensors with different ER heights. The magnitude of f_{ER} increased by about 3.0 dB as the ER height increased from 25 μm to 75 μm, demonstrating that the increased height of the ER had an amplifying effect for shear sensing due to the increased torque. However, the magnitude of

f_{ER} was greatly decreased at 100 μm in ER height. It seems that there was a limit to how high the ER could be made to take advantage of the amplifying effect, since beyond a critical ER height, or aspect ratio, we believe the ER structures may buckle and absorb the shear forces, leading to reduced force transfer. This height-dependent degradation has also been observed in other reports [30].

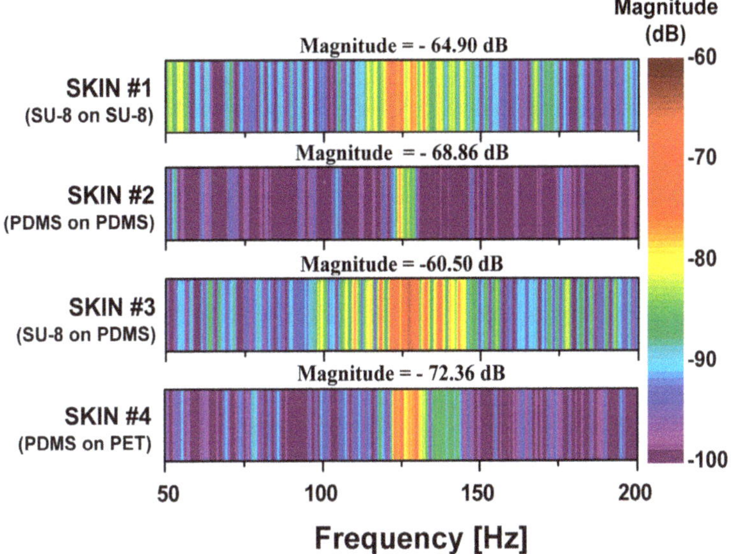

Figure 6. FFT results of tactile sensors with four types of SKINs when a PET tip was scanned at 2.5 cm/s on the tactile sensor. The indicated magnitude values show the magnitude of peak frequency (f_{ER}) induced by a 200-μm period of ERs and a 2.5 cm/s scanning speed.

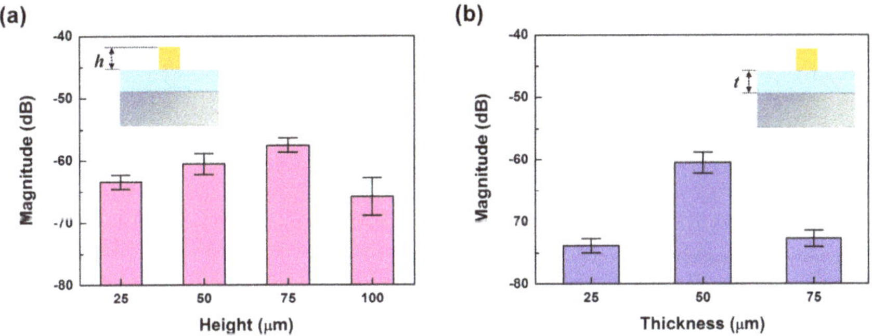

Figure 7. Dependence of the magnitude of f_{ER} on (**a**) the height (h) of the ER and (**b**) the thickness (t) of the epidermal board.

We also investigated the relationship between the tactile sensitivity and the thickness of the epidermal board. We used the same material configuration of SKIN #3 at a scanning speed of 2.5 cm/s, but with different board thicknesses varying between 25, 50, and 75 μm, and kept the height of the ER at 50 μm. We found that the highest magnitude of f_{ER} in the tactile sensor was from the 50-μm-thick epidermal board. This indicates that there may be an optimum thickness for the epidermal board that will give the highest force transfer while reducing the dispersion of the vertical force [31]. Due to these results, we chose SKIN #3 with an ER height of 75 μm and an epidermal board 50 μm thick for the following texture sensing.

5. Surface Period Detection

We evaluated the surface period-sensing characteristics of the developed tactile sensor using two periodic grating structures with a 150-μm and 625-μm period (insets of Figure 8a,b). The periodic grating structures were fabricated by a patterned 60-μm-thick SU-8. Figure 8a,b shows the FFT results of tactile sensor output induced by scanning performed at 1–4 mm/s. When the grating structure slid on the tactile sensor, individual ERs periodically interacted with each grating structure, and simultaneously the individual grating structure periodically interacted with each ER. Therefore, the FFT results present the f_{ER} and the peak induced by a surface period of a contact object. We observed that the distribution of the peaks showed a blue shift as the scanning speed increased, with the scanning speed divided by the highest peak frequency remaining the same. At a 2-mm/s scanning speed (Figure 3a), a distinct peak for f_{ER} was observed at 10 Hz, and peaks at 13.3 Hz were induced by the 150-μm period of the surface grating structure, with harmonic signals appearing at each 10- and 13.3-Hz multiple. We saw similar characteristics for the 625-μm periodic structure. This showed that our sensor was able to detect surface structures under 100 μm (75 μm wide and 60 μm high). Since the highest peak would come from the ER's interaction with the contacting surface, the f_{ER} offered information on scanning speed. Then the other peaks in the spectra could elucidate the surface periods of the object with the known scanning speed.

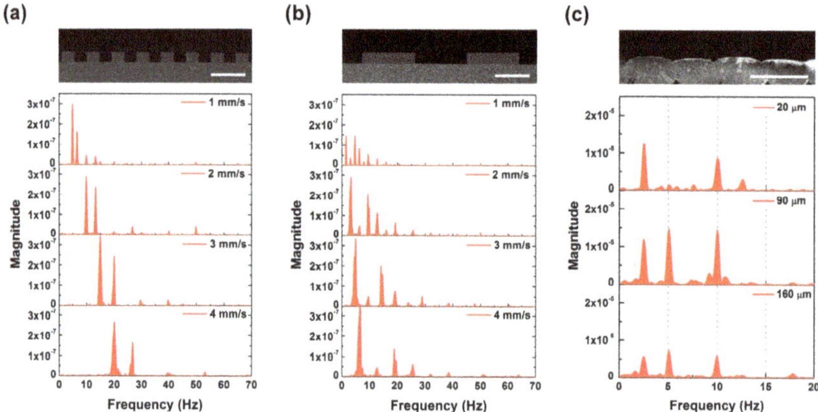

Figure 8. Surface period-detecting characteristics of biomimetic tactile sensor: the FFT results of the tactile sensor output induced by scanning the grating structures with (**a**) a 150-μm and (**b**) 625-μm period at 1–4 mm/s scanning speed, where the insets show cross-sectional SEM images of the grating structures (scale bars indicate 200 μm); (**c**) the FFT results of the tactile sensor output induced by scanning a 3D-printed structure at 2 mm/s (varying the contact depth), where the inset shows a cross-sectional SEM image of the 3D-printed structure (scale bar indicates 500 μm).

In sensing the surface roughness, varying the contact depth was an important factor in measuring the surface structure with different heights. A plastic (polylactic acid) surface with a 390-μm-period elliptic structure with a 20-μm height difference was fabricated by using a fused filament fabrication-type 3D printer, as seen in the inset of Figure 8c. The reciprocating direction of the printing nozzle resulted in the height difference. Figure 8c shows the FFT results of the tactile sensor when the printing structure was scanned at 2 mm/s, increasing the contact depth. Comparing the FFT result at a 90-μm contact depth to that at 20 μm, we observed an additional peak at 5.12 Hz, reflecting the period of the printing structure with a lower height. The peak at 5.12 Hz was not considered to be a harmonic of 2.56 Hz due to its higher magnitude. This result shows that our tactile sensor was able to detect not only the surface period but also a 20-μm height difference in surface structure. In the deeper contact depth of 160 μm, the signal magnitude of FFT decreased. It seems that the ER was not able to deform and recover fast enough due to the increased pressure resulting from the increased contact depth.

When the SKIN interacts with the object's surface, the generated vibration should depend on the frictional aspect of the materials involved. Therefore, even with the same surface roughness, if the material is different, it should produce vibrations with different spectral distributions. We fabricated a PDMS replica of a finger, which should have had identical surface structures to the actual finger used, and measured the sensor signal through tactile scanning to see how the measured signals would differ. Figure 9a,b shows the FFT results, and it is clear that the two results were quite different and distinguishable. In both cases, distinct peaks were observed identically at f_{ER} = 10 Hz (representing the ER period) and at 4.2 Hz (which was dependent on the 480-μm period of the human fingerprint). However, differences in the amplitudes of major peaks and their harmonics and spectral distribution in the lower frequency region were observed. This difference seems to have been caused by the stiffness difference between the human finger and the PDMS replica. The human finger showed larger deformability due to its Young's modulus (~100 kPa [25]) being lower than PDMS (750 kPa). The human skin's lower modulus led to a larger contact area and higher friction with the SKIN layer of the sensor. This allowed the characteristic peaks to become distributed, forming a higher amplitude spectrum in the low-frequency region. In addition, the chemical characteristics of a finger surface (cell membrane, oil, and sweat) differ from those of PDMS, which produces differing frictional characteristics. This result showed that our sensor was able to distinguish between different materials even if they had the same surface structure.

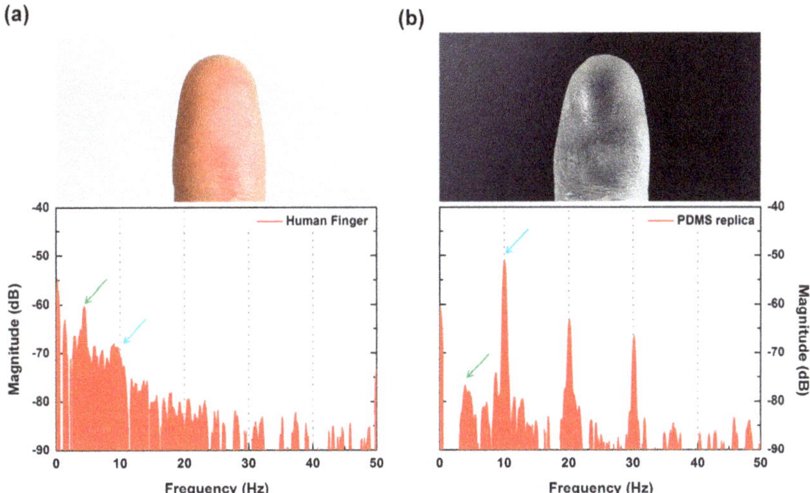

Figure 9. FFT results of tactile sensor output induced by scanning (**a**) a human finger and (**b**) a PDMS replica, where the insets show optical images of a human finger pad and the PDMS replica. Blue arrows indicate the f_{ER}, and green arrows indicate the peaks of ERs of the contact objects.

6. Texture Detection

The ability to discern surface feature periods and contact materials may be used to distinguish the characteristics of similar materials. We performed surface scanning measurements on similar materials with different compositions to test this ability of our biomimetic tactile sensor. First, we compared the scanning results between a sheet of printing paper and a sheet of papyrus at a 2 mm/s scanning speed (see Figure 10a,b). The results taken from the printing paper scanning showed that the peaks at f_{ER} and its integral multiples showed far lower amplitudes than did papyrus. These complex spectra could be simply analyzed using the energy spectrum density (ESD), which is the integral of FFT in a specific frequency range. Comparing the two results in the frequency range under and over f_{ER}, the ESDs of the papyrus were 22 times and 13 times larger than those of the printing paper. It can be said that the papyrus had more surface features with periods shorter and longer than 200 μm. This can be verified visually (as shown in the optical and SEM images of Figure 10a,b), as the printing paper was featureless while the papyrus was feature-rich with many observable line structures. The surface features of the papyrus had higher directionality, periodicity, and height than did the printing paper. Due to these structural differences in surface topology, papyrus had the higher peak amplitude over most of the observed frequency range. Since our fingertips felt that the papyrus surface was "rougher" than that of the printing paper, the higher peaks in the FFT results may reflect the roughness of the material surface.

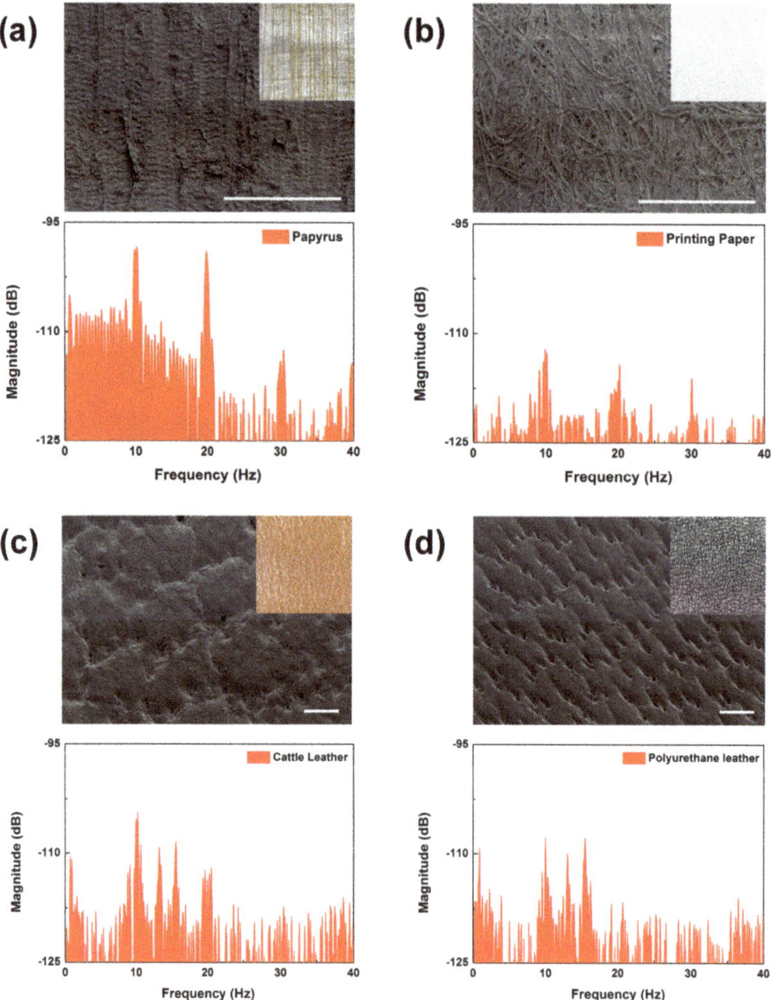

Figure 10. FFT results of tactile sensor output induced by scanning (at a 2-mm/s scanning speed) (**a**) a sheet of papyrus, (**b**) a sheet of printing paper, (**c**) cattle leather, and (**d**) polyurethane artificial leather, where insets show SEM and optical images of each contact object (scale bars indicate 500 μm).

For real leather and artificial leather, it is quite difficult for an untrained person to distinguish between the two just by scanning their fingers over the surfaces. We found that the scanning results of these materials did produce FFT results that were discernable. As shown in the inset images of Figure 10c,d, cattle leather and polyurethane leather have similar surface structures with many grains and pores. The FFT results showed a similar spectral distribution, which made it understandable why it would be difficult to distinguish between them just by relying on touch. However, there were features that stood out. In the frequency range from 8 to 22 Hz, the ESD of the cattle leather was 2 times larger than that of the polyurethane leather. This shows that our sensor may be used to detect fine differences between artificial and real leather.

The results of Figure 8c show that the contact depth was an essential factor in detecting more detailed surface characteristics. Varying the contact depth, we measured the change in the frequency spectrum between two fabrics with different hardness. As seen in the optical image of Figure 11a,b, the harder fabric had tight weaves and the softer fabric had hair-like structures. The topmost spectrum was recorded when the 10-Hz fingerprint peaks started to appear. Then the fabrics were scanned with the sensor pressed on its surface with the pressure depth increasing in 20-μm steps toward a 200-μm total depth. In the low-contact depth, we observed similar frequency spectra, which showed no specific peak except for f_{ER} despite the significant visual differences between the two fabrics. Increasing the contact depth, we observed the appearance of peaks at 1.6 Hz and 0.2 Hz and their harmonics with a 1.2-mm period of the hard fabric and a 10-mm period of the soft fabric. Comparing the two FFT results, the harder-fabric FFT showed more distinct peaks appearing as the scanning pressure increased to that of the softer-fabric FFT. We can conclude that the frequency spectra taken at various contact depths can give additional hardness information, which will aid in increasing the ability of the biomimetic tactile sensor to distinguish between various materials and which may make tactile sensor-based material identification a reality.

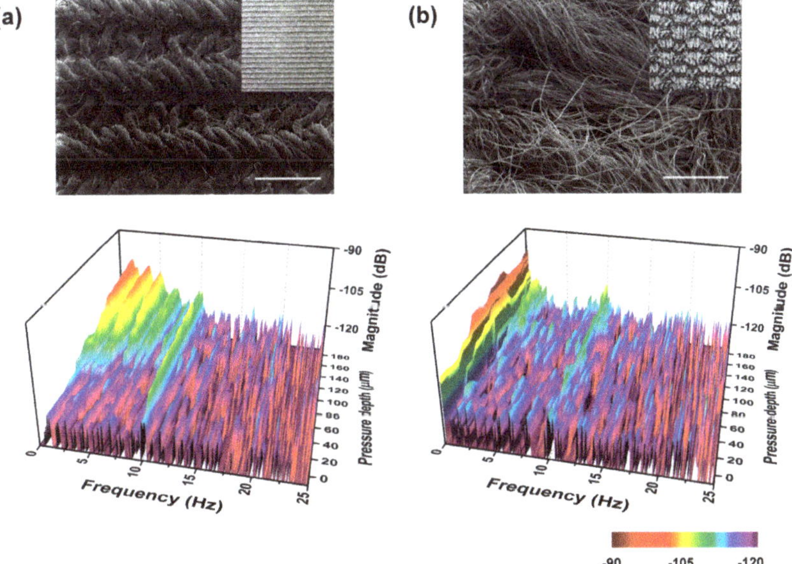

Figure 11. FFT results of tactile sensor output induced by scanning (**a**) a hard fabric with tight weaves and (**b**) a soft fabric with hair-like structures, increasing the contact depth. Inset images show SEM and optical images of the two different fabrics (scale bars indicate 1 mm).

7. Discussion

Previously reported tactile sensors that were based on measuring scanning vibrations commonly used accelerometers and microphones [32,33]. These sensors were built into or on top of an artificial finger, and vibration damping could occur, which would reduce the accuracy of the results. Since our biomimetic tactile sensor would be in direct contact with an object, it could effectively detect the vibration information induced by the interaction between the ERs and the contact object without damping. Our SKIN showed tactile sensitivity that was high enough to detect sub-100-μm surface structures of a sheet of papyrus. It has been reported that ERs show good tactile sensitivity on surfaces with structural periods that are 0.5–2 times as long as that of the ERs [26]. When varying the design of the ER period, the tactile sensitivity may be adjusted to the target sensitivity, which may be required for various applications.

Conventional fingerprint identification security systems can detect the surface structure of human finger pads but cannot discriminate between finger pad replicas. In the results in Figure 9, it is shown that our biomimetic tactile sensor had the capability of discriminating between a human finger pad and its replica, which may make them applicable for future security systems. By utilizing its mechanical flexibility, if we incorporate the biomimetic tactile sensor into the previously reported tactile stiffness sensor [34], it may be possible to develop a portable tactile measurement system that can detect surface roughness as well as stiffness, which will make tactile surface recognition possible.

8. Conclusions

We developed a SKIN with a bilayer structure with different elastic moduli that mimics human ER and epidermis characteristics. When a hard ER and a soft epidermal board were used for the SKIN, we improved the effectiveness of transferring surface shear forces into vertical vibrations. The SKIN with a hard SU-8 ER and a soft PDMS epidermal board showed 6.85 times higher tactile sensitivity than did the one fabricated using only the soft PDMS. We optimized the ER dimensions and the epidermal board thickness, and the biomimetic tactile sensor with the optimized SKIN was able to detect a 75-μm period and a 20-μm height difference in a contact surface. Our sensor also showed an ability to sense the difference between contact with a human finger and a PDMS replica with the same surface structure. In addition, we demonstrated that it was possible to discern the differences in textures of paper, leather, and fabric using the biomimetic tactile sensor. With further development and through use of the sensor's mechanical flexibility, our biomimetic tactile sensor may make texture-recognizing artificial fingers a possibility.

Author Contributions: conceptualization, E.C.; formal analysis, E.C., O.S., H.S., and S.-B.L.; funding acquisition, S.-B.L.; investigation, S.K., S.Y., G.R., and Y.S.; methodology, E.C., O.S., and S.-B.L.; project administration, S.-B.L.; resources, J.L., Y.S.; supervision, S.-B.L.; visualization, H.Y. and O.S; writing—original draft, E.C.; writing—review and editing, S.-B.L.

Funding: This research was funded by the Basic Science Research Program through the National Research Foundation of Korea (NRF) (funded by the Ministry of Education, Science, and Technology (2012R1A6A1029029) and the Ministry of Education (2019R1I1A1A01059713)) and partly by the Samsung Research Funding Center of Samsung Electronics under Project Number SRFC-IT1701-03.

Conflicts of Interest: The authors declare no conflict of interest.

References

1. Kuilenburg, J.; Masen, M.A.; Heide, E. A review of fingerpad contact mechanics and friction and how this affects tactile perception. *Proc. IMechE Part J J. Eng. Tribol.* **2013**, *229*, 243–258. [CrossRef]
2. Pang, G.; Deng, J.; Wang, F.; Zhang, J.; Pang, Z.; Yang, G. Development of flexible robot skin for safe and natural human-robot collaboration. *Micromachines* **2018**, *9*, 576. [CrossRef] [PubMed]
3. Kaboli, M.; Cheng, G. Robust tactile descriptors for discriminating objects from textural properties via artificial robotic skin. *IEEE Trans. Robot.* **2018**, *34*, 985–1003. [CrossRef]

4. Odhner, L.U.; Jentoft, L.P.; Claffee, M.R.; Corson, N.; Tenzer, Y.; Ma, R.R.; Buehler, M.; Kohout, R.; Howe, R.D.; Dollar, A.M. A compliant, underactuated hand for robust manipulation. *Int. J. Robot. Res.* **2014**, *33*, 736–752. [CrossRef]
5. Girão, P.S.; Ramos, P.M.P.; Postolache, O.; Pereira, J.M.D. Tactile sensors for robotic applications. *Measurement* **2013**, *46*, 1257–1271. [CrossRef]
6. Wang, S.; Xu, J.; Wang, W.; Wang, G.J.N.; Rastak, R.; Molina-Lopez, F.; Chung, J.W.; Niu, S.; Feig, V.R.; Lopez, J.; et al. Skin electronics from scalable fabrication of an intrinsically stretchable transistor array. *Nature* **2018**, *555*, 83–88. [CrossRef] [PubMed]
7. Kim, J.; Lee, M.; Shim, H.J.; Ghaffari, R.; Cho, H.R.; Son, D.; Jung, Y.H.; Soh, M.; Choi, C.; Jung, S.; et al. Stretchable silicon nanoribbon electronics for skin prosthesis. *Nat. Commun.* **2014**, *5*, 1–11. [CrossRef]
8. Kim, D.-H.; Lu, N.; Ma, R.; Kim, Y.-S.; Kim, R.-H.; Wang, S.; Wu, J.; Won, S.M.; Tao, H.; Islam, A.; et al. Epidermal electronics. *Science* **2011**, *333*, 838–843. [CrossRef]
9. Zhang, J.; Zhou, L.J.; Zhang, H.M.; Zhao, Z.X.; Dong, S.L.; Wei, S.; Zhao, J.; Wang, Z.L.; Guo, B.; Hu, P.A. Highly sensitive flexible three-axis tactile sensors based on the interface contact resistance of microstructured graphene. *Nanoscale* **2018**, *10*, 7387–7395. [CrossRef]
10. Wu, Y.; Liu, Y.; Zhou, Y.; Man, Q.; Hu, C.; Asghar, W.; Li, F.; Yu, Z.; Shang, J.; Liu, G.; et al. A skin-inspired tactile sensor for smart prosthetics. *Sci. Robot.* **2018**, *3*, eaat0429.
11. Kim, S.Y.; Park, S.; Park, H.W.; Park, D.H.; Jeong, Y.; Kim, D.H. Highly sensitive and multimodal all-carbon skin sensors capable of simultaneously detecting tactile and biological stimuli. *Adv. Mater.* **2015**, *27*, 4178–4185. [CrossRef] [PubMed]
12. Viry, L.; Levi, A.; Totaro, M.; Mondini, A.; Mattoli, V.; Mazzolai, B.; Beccai, L. Flexible three-axial force sensor for soft and highly sensitive artificial touch. *Adv. Mater.* **2014**, *26*, 2659–2664. [CrossRef] [PubMed]
13. Kang, D.; Pikhitsa1, P.V.; Choi, Y.W.; Lee, C.; Shin, S.S.; Piao, L.; Park, B.; Suh, K.-Y.; Kim, T.; Choi, M. Ultrasensitive mechanical crack-based sensor inspired by the spider sensory system. *Nature* **2014**, *516*, 222–226. [CrossRef] [PubMed]
14. Wang, X.; Gu, Y.; Xiong, Z.; Cui, Z.; Zhang, T. Silk-molded flexible, ultrasensitive, and highly stable electronic skin for monitoring human physiological signals. *Adv. Mater.* **2014**, *26*, 1336–1342. [CrossRef] [PubMed]
15. Mannsfeld, S.C.B.; Tee, B.C.-K.; Stoltenberg, R.M.; Chen, C.V.H.-H.; Barman, S.; Muir, B.V.O.; Sokolov, A.N.; Reese, C.; Bao, Z. Highly sensitive flexible pressure sensors with microstructured rubber dielectric layers. *Nat. Mater.* **2010**, *9*, 859–864. [CrossRef] [PubMed]
16. Bensmaia, S.; Hollins, M. Pacinian representations of fine surface texture. *Percept. Phychophysics* **2005**, *67*, 842–854. [CrossRef] [PubMed]
17. Scheibert, J.; Leurent, S.; Prevost, A.; Debrégeas, G. The role of fingerprints in the coding of tactile information probed with a biomimetic sensor. *Science* **2009**, *323*, 1503–1506. [CrossRef]
18. Oddo, C.M.; Beccai, L.; Wessberg, J.; Wasling, H.B.; Mattioli, F.; Carrozza, M.C. Roughness encoding in human and biomimetic artificial touch: Spatiotemporal frequency modulation and structural. *Sensors* **2011**, *11*, 5596–5615. [CrossRef]
19. Ding, S.; Pan, Y.; Tong, M.; Zhao, X. Tactile perception of roughness and hardness to discriminate materials by friction-induced vibration. *Sensors* **2017**, *17*, 2748. [CrossRef]
20. Koc, I.M.; Aksu, C. Tactile sensing of constructional differences in fabrics with a polymeric fingertip. *Tribol. Int.* **2013**, *59*, 339–349. [CrossRef]
21. Loeb, G.E.; Fishel, J.A. Bayesian exploration for intelligent identification of textures. *Front. Neurorobot.* **2012**, *6*, 4. [CrossRef]
22. Delmas, P.; Hao, J.; Rodat-Despoix, L. Molecular mechanisms of mechanotransduction in mammalian sensory neurons. *Nat. Rev. Neurosci.* **2011**, *12*, 139–153. [CrossRef] [PubMed]
23. Wickett, R.R.; Visscher, M.O.; Ohio, C. Structure and function of the epidermal barrier. *Am. J. Infect. Control* **2006**, *34*, S98–S110. [CrossRef]
24. Hollins, M.; Bensmaia, S.J.; Washburn, S. Vibrotactile adaptation impairs discrimination of fine, but not coarse, textures. *Somatosens. Motor Res.* **2001**, *18*, 253–262. [CrossRef]
25. Liang, X.; Boppart, S.A. Biomechanical properties of in vivo human skin from dynamic optical coherence elastography. *IEEE Transcations Biomed. Eng.* **2010**, *57*, 953–959. [CrossRef] [PubMed]
26. Fagiani, R.; Massi, F.; Chatelet, E.; Berthier, Y.; Akay, A. Tactile perception by friction induced vibration. *Tribol. Int.* **2011**, *44*, 1100–1110. [CrossRef]

27. Fernández-Blázquez, J.P.; Fell, D.; Bonaccurso, E.; Campo, A. Superhydrophilic and superhydrophobic nanostructured surfaces via plasma treatment. *J. Colloid Interface Sci.* **2011**, *357*, 234–238. [CrossRef] [PubMed]
28. Choi, E.; Sul, O.; Kim, J.; Kim, K.; Kim, J.-S.; Kwon, D.-Y.; Choi, B.-D.; Lee, S.-B. Contact pressure level indication using stepped output tactile sensors. *Sensors* **2016**, *16*, 511. [CrossRef]
29. Ephraim, Y.; Malah, D. Speech enhancement using a minimum-mean square error short-time spectral amplitude estimator. *IEEE Trans. Acoust. Speech Signal Process.* **1984**, *32*, 1109–1121. [CrossRef]
30. Zhang, Y. Sensitivity enhancement of a micro-scale biomimetic tactile sensor with epidermal ridges. *J. Micromech. Microeng.* **2010**, *20*, 085012. [CrossRef]
31. Aprilia, L.; Nuryadi, R.; Hartanto, D. Sensitive layer thickness dependence on microcantilever sensor sensitivity. *Adv. Mater. Res.* **2013**, *789*, 219–224. [CrossRef]
32. Koiva, R.; Schwank, T.; Walck, G.; Haschke, R.; Ritter, H.J. Mechatronic fingernail with static and dynamic force sensing. In Proceedings of the 2018 IEEE/RSJ International Conference on Intelligent Robots and Systems (IROS), Madrid, Spain, 1–5 October 2018.
33. Edwards, J.; Lawry, J.; Rossiter, J.; Melhuish, C. Extracting textural features from tactile sensors. *Bioinsp. Biomim.* **2008**, *3*, 035002. [CrossRef] [PubMed]
34. Sul, O.; Choi, E.; Lee, S.B. A portable stiffness measurement system. *Sensors* **2017**, *17*, 2686. [CrossRef] [PubMed]

 *micromachines*

Article

Flexible Tactile Sensor Array for Slippage and Grooved Surface Recognition in Sliding Movement

Yancheng Wang [1,2,*], Jianing Chen [2] and Deqing Mei [1,2]

1 State Key Laboratory of Fluid Power and Mechatronic Systems, School of Mechanical Engineering, Zhejiang University, Hangzhou 310027, China
2 Key Laboratory of Advanced Manufacturing Technology of Zhejiang Province, School of Mechanical Engineering, Zhejiang University, Hangzhou 310027, China
* Correspondence: yanchwang@zju.edu.cn

Received: 16 July 2019; Accepted: 28 August 2019; Published: 30 August 2019

Abstract: Flexible tactile sensor with contact force sensing and surface texture recognition abilities is crucial for robotic dexterous grasping and manipulation in daily usage. Different from force sensing, surface texture discrimination is more challenging in the development of tactile sensors because of limited discriminative information. This paper presents a novel method using the finite element modeling (FEM) and phase delay algorithm to investigate the flexible tactile sensor array for slippage and grooved surfaces discrimination when sliding over an object. For FEM modeling, a 3×3 tactile sensor array with a multi-layer structure is utilized. For sensor array sliding over a plate surface, the initial slippage occurrence can be identified by sudden changes in normal forces based on wavelet transform analysis. For the sensor array sliding over pre-defined grooved surfaces, an algorithm based on phase delay between different sensing units is established and then utilized to discriminate between periodic roughness and the inclined angle of the grooved surfaces. Results show that the proposed tactile sensor array and surface texture recognition method is anticipated to be useful in applications involving human-robotic interactions.

Keywords: finite element modeling; surface texture; grooved surface; tactile sensor array; wavelet transform; spectral analysis; inclined angle

1. Introduction

Flexible tactile sensors have been widely utilized in robotics, prosthetic hands, and medical surgery [1,2]. For grasping and manipulation tasks, the robotic hand with integrated tactile sensors can perceive tactile information between the hand, fingers, and grasped objects. This tactile information plays an important role and can be used for robotic feedback control [3,4]. For daily grasping in robotic and prosthetic hands, if the applied grasping force is too low, objects may slip through the hand, while fragile objects may be damaged when the applied force is too large. Furthermore, the roughness, texture, material hardness, and contour of the objects also affect the requisite grasping force [5]. Therefore, robotic dexterous manipulation generally requires integrated tactile sensors on the robotic hand with force sensing as well as object texture and contour shape recognition abilities.

The tactile sensor array is usually designed with several sensing units arranged in a row/column configuration and can be used to measure distributed contact forces [6,7]. In the past decade, developments in the tactile sensor array have attracted many researchers, and several types of tactile sensor array have been proposed [8–10]. Recently, we utilized conductive rubber as the sensing material to develop a flexible tactile sensor array with 3×3 sensing units which can be worn on the finger of a robotic hand and can measure three-axis contact forces during grasping applications [11,12]. The contact behavior of the sensor array with objects affects the contact force sensing performance of the tactile sensor array we developed. Analytical modeling was conducted to study the sensing performance and mechanical behavior in many researches. The basic structure of the tactile sensor is usually first simplified into a combination of cylinders, cuboids and other basic geometries. Then a lumped parameter model can be developed to analyze the mechanical properties of the tactile sensor. Zhang et al. [13] utilized the Stribeck friction model to study the mechanical behavior between a rigid gripper and the gripped object during the initial slippage phase. They found that the induced normal force changes suddenly when slipping occurs due to change in the static/dynamic friction coefficient. Ho et al. [14,15] simplified the fingertip into a bundle of beam to calculate localized displacement for slip detection.

For a tactile sensor array with a more complex structural design, numerical modeling will be an effective approach to study the performance and contact behavior of the tactile sensors. Dao et al. [16] presented a numerical model in Marc Mentat software that analyzed the normal stress distribution of the sensing units when an external force is applied and identified the optimal location for the arrangement of piezoresistors. Youssefian et al. [17] developed a finite element modeling (FEM) of the tactile sensor by adopting nonlinear elastic material properties to study the induced stress and strain when a normal force is applied to the outer surface of the tactile sensors. In these proposed numerical models, the structure of the tactile sensors needs to be simplified into beam and plate structures for fast calculation convergence. Therefore, this will inevitably affect the accuracy of the mechanical behaviors of the tactile sensors for external force sensing, like the filtering effects of the sensor's top cover material is neglected [18]. Thus, to analyze the sensing performance and contact behavior, an accurate 3D FEM model of the tactile sensor array needs to be developed, and this is a goal of this research.

As mentioned earlier, the surface roughness, hardness, texture and contour shape of the object affect the sensing performance of the tactile sensor array. For surface texture recognition, two approaches have been validated to have the ability to extract the object's features. (1) Using a tactile sensor array with high-density sensing units to measure the contact forces when it touches the object's surface. Then the measured force values are plotted into a gray scale figure, which can be used to discriminate between the contour shapes of the objects using an image processing algorithm [19,20]. (2) Using a spectral analysis algorithm to analyze the measured forces when the tactile sensor slides along the surface of the objects. Oddo et al. [21] utilized a 2 × 2 tactile sensor array to measure the normal forces when sliding over the patterned surfaces, and fast Fourier transformation (FFT) was used to discriminate between the surface roughness and periodic information from the grooved surfaces. Further, they developed an approach using a machine learning algorithm (k-NN classifier) and wavelet transform to classify the surface texture [22]. In 2012, Fishel et al. [23] utilized a biologically inspired tactile sensor (BioTac) to measure tactile vibrations and reaction forces when exploring surfaces with different textures. The Bayesian exploration algorithm was then used to analyze the force data obtained, and 117 types of textures were successfully identified. This algorithm requires plenty of input data for the improvement of accuracy, and this limits the application of the BioTac sensor for surface texture recognition. Therefore, based on the obtained reaction forces of the tactile sensor array, an effective surface texture recognition method still needs to be developed.

Therefore, the proposed tactile sensor for surface texture recognition still needs to be investigated, the present surface recognition method can still not be used for practical usage. To fill this research gap, we developed an accurate 3D FEM model of the tactile sensor array to study the sensing performance and contact behavior of the sensor when contacted with objects. Based on the measured contact forces of the tactile sensor array, a novel approach based on phase delay algorithm for grooved surface recognition is developed and verified by both FEM modeling and experimental validation. The main content of this paper is divided as follow: Section 2 presents the structure and working principles of the tactile sensor array on which 3D FEM modeling was conducted. Section 3 describes the experimental setup and procedures. Two sets of experiments were conducted: slippage detection and surface texture recognition when the sensor array slides over the object's surface. The FEM simulation, experimental results, and discussion are presented in Section 4.

2. Design of Tactile Sensor Array and FEM Modeling

2.1. Flexible Tactile Sensor Array

Of the tactile sensing principles, piezoresistive sensing is selected because of its relatively simple structural design and good anti-noise performance. Highly sensitive INASTAMOR pressure conductive rubber (from Inaba Rubber Co. Ltd., Osaka, Japan) is utilized as the sensitive material and cut into small pieces of round-shaped chips with a diameter of 3.0 mm. The structural design of the 3 × 3 flexible tactile sensor array is illustrated in Figure 1a,b. This sensor array mainly consists of three layers: top polydimethylsiloxane (PDMS) bump, a middle room temperature vulcanizable (RTV) adhesive layer with conductive rubber chips, and bottom electrodes on a thin film of polyethylene terephthalate (PET). The thicknesses of these three layers are 0.8, 0.5 and 0.1 mm, respectively. The distance between adjacent units is about 3.5 mm, and thus the overall dimensions of the tactile sensor array are 20 mm × 16 mm × 1.4 mm. A detailed structural design of the flexible tactile sensor array can be found in one of the references [12].

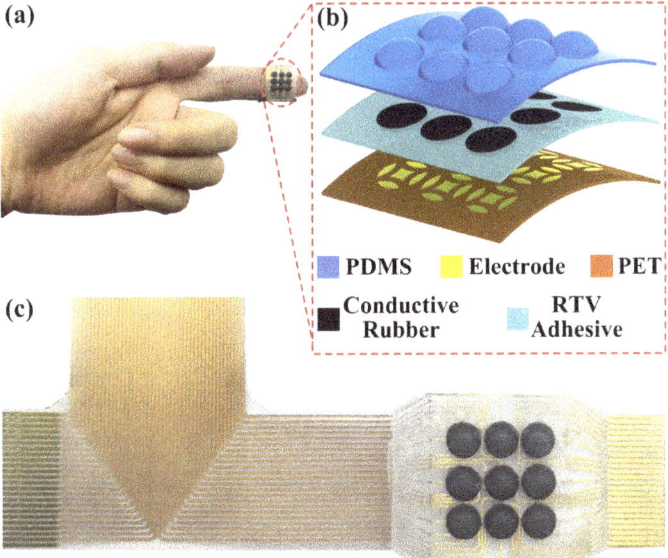

Figure 1. (**a**,**b**) Structure of the flexible tactile sensor array; (**c**) Fabricated tactile sensor array.

The patterned electrodes underneath the rubber chip have four side electrodes and one central common electrode, which generates four resistors (R_1, R_2, R_3 and R_4) and divides the sensing unit into five areas, as shown in Figure 1b. Thus, these four resistors can measure the changes in resistance for external three-axis force sensing. Typically, as the tactile sensor array is worn on the finger of a hand for grasping and touching objects, the external force will be exerted over the sensor array, and the induced deformation of the PDMS bump and conductive rubber chips will change the resistances of these four resistors.

2.2. FEM Modeling

For FEM modeling, the finite element mesh of the 3D tactile sensor array model with dimensions of 20 mm × 16 mm × 1.4 mm is shown in Figure 2a. The accurate 3D geometry of the bump, rubber chip, and substrate film layers were converted to the FEM mesh using ABAQUS (v6.14, Dassault Systèmes Simulia Corp., Providence, RI, USA). Both "structured" and "sweep" algorithms for element mesh were utilized. The bump layer, rubber chip and surrounding RTV adhesive, and the PET film were connected using the "tie" function to lock the nodes onto the surfaces. To ensure perfectly tied surfaces, the mesh (node positions) on the mating surface must be consistent. For mesh convergence, the region of the bump and rubber chip and other contact regions were finely meshed, as shown in Figure 2b. Altogether, the tactile sensor array was meshed using 78,885 eight-node hexahedron elements. The element numbers in each layer are as follows: bump (53,236 elements), conductive rubber chip (10,368 elements), RTV adhesive (14,001 elements) and bottom PET film (1280 elements). For boundary conditions, the PDMS bump, conductive rubber chip, RTV adhesive, and PET film layers were merged together. The underside of the PET layer was confined by the displacement boundary condition.

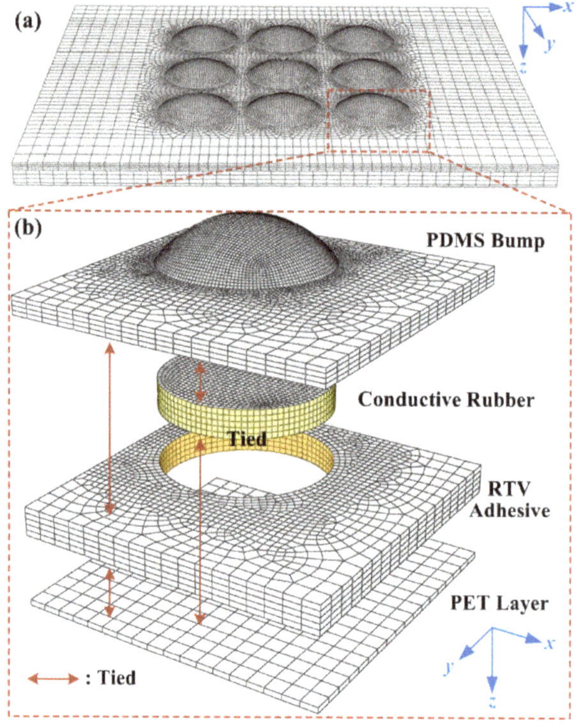

Figure 2. (**a**) 3D finite element modeling (FEM) model of flexible tactile sensor array; (**b**) Close-up view of sensing unit.

The mechanical properties of the PET film were adopted from a previous study while Young's modulus and Poisson ratio are about 3000 MPa and 0.47 [24]. For the conductive rubber, PDMS and RTV adhesive, uniaxial compression tests were conducted according to American Society for Testing and Materials (ASTM) standards [25]. The measured nominal stress versus nominal strain curves for the rubber, PDMS and RTV adhesive materials are shown in Figure 3.

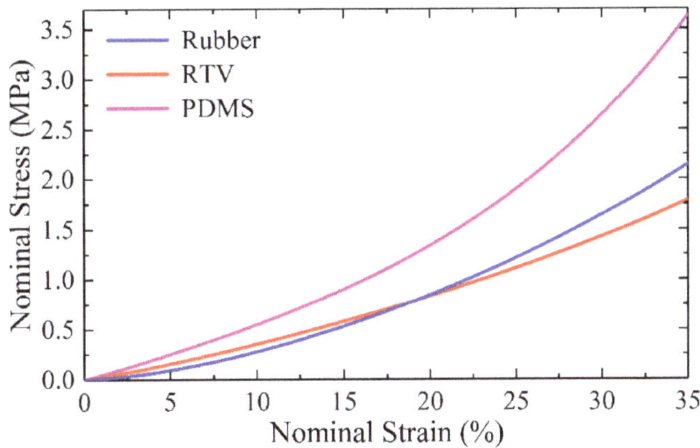

Figure 3. Measured nominal stress versus strain curves of rubber, room temperature vulcanizable (RTV) and polydimethylsiloxane (PDMS) materials.

All three stress-strain curves have nonlinear elastic behaviors, especially under large strains. The hyper-elastic Yeoh model [26] was used in the ABAQUS software to represent the nonlinear properties of the conductive rubber, PDMS and RTV adhesive materials. For these incompressible materials, the strain energy density function can be expressed as

$$W(I_1) = C_1(I_1 - 3) + C_2(I_1 - 3)^2 + C_3(I_1 - 3)^3 \quad (1)$$

where I_1 stands for the first invariant of the Green deformation tensor and C_i is the material parameter. Under the circumstance of uniaxial compression, Equation (1) can be transformed into the form that describes the relation between the stress σ and strain ε as

$$\begin{aligned} \sigma = & 2C_1(\lambda - \lambda^{-2}) + 4C_2(\lambda^3 - 3\lambda + 1 + 3\lambda^{-2} - 2\lambda^{-3}) \\ & +6C_3(\lambda^5 - 6\lambda^3 + 3\lambda^2 + 9\lambda - 6 - 9\lambda^{-2} + 12\lambda^{-3} - 4\lambda^{-4}) \end{aligned} \quad (2)$$

where λ is the elongation and equals to 1 + ε.

By using the least squares fitting, three parameters (C_1, C_2, and C_3) in the Yeoh model for rubber, RTV adhesive, and PDMS are obtained, and these are listed in Table 1.

Table 1. Material properties of rubber, RTV, and PDMS materials.

Material	**Yeoh Model**			**Poisson Ratio**
	C_1	C_2	C_3	
Conductive rubber	0.5686	0.0540	−0.0181	0.47
RTV adhesive	0.5551	−0.0356	0.0027	0.48
PDMS	0.7997	0.2881	−0.0375	0.47 [27]

3. Experimental Setup and Procedure

3.1. Experimental Setup

The entire experimental setup is shown in Figure 4. It mainly consists of an *xyz* linear motion stage and a three-axis commercialized force sensor. The developed flexible tactile sensor array was attached to a plastic loading bar, which was mounted to a *z*-axis motion stage. A scanning circuit based on digital signal processing (TMS320F2812, Texas Instruments Inc, Dallas, TX, USA) was designed and used for the distributed normal and shear forces measurement [12]. During the experiments, two types of surfaces (flat surface and grooved surface) made using stereolithography (SLA) technique were utilized. The flat surface was used for slippage detection when the loading bar and tactile sensor array slid on the surface. The grooved surfaces with different patterns were used for surface texture recognition.

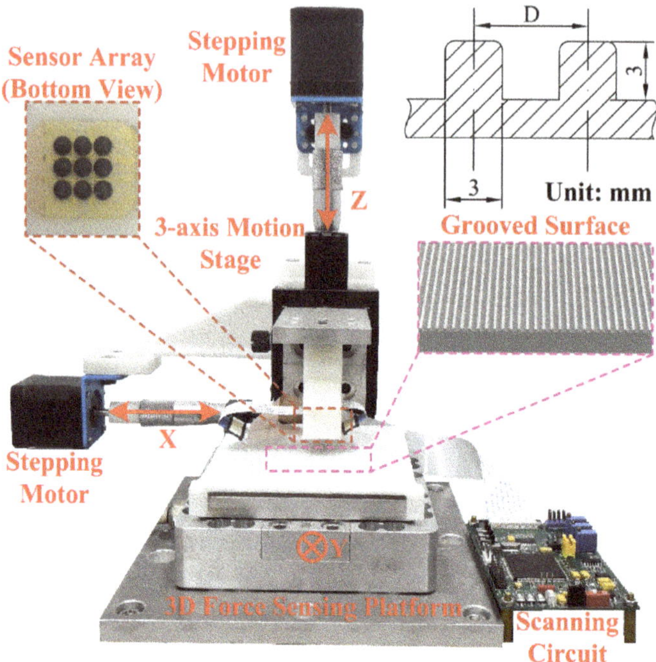

Figure 4. Experimental setup for the validation tests.

3.2. Experimental Procedure

During the experiments, the motion speed of the linear stage was controlled by the stepping motors. To reduce the inertia effects, the sliding movement of the loading bar and tactile sensor array should be lower than 1.0 mm/s. For experimental validation, two sets of experiments were conducted:

Slippage detection. First, the loading bar with a tactile sensor array compressed the flat surface for 8 s, and the induced compression force was increased up to 20 N. This force was a little large than that of FEM simulation, because this force is sufficient to overcome the effect of the death zone, and make the output voltage of each sensing unit clear enough (over 0.5 V) for further analyses. Secondly, this is followed by a holding stage that lasted for 15 s. Thirdly, the loading bar and tactile sensor array slid along the plate surface for about 25 s at a constant speed of 0.25 mm/s.

Surface texture recognition. Three grooved surfaces with different spatial periods of 0.9, 1.2, and 1.5 mm were utilized. The inclined angle (α), defined as the angle between the grooves and the y-axis, was set as 0–60° with an increment of 15°. The tactile sensor array first compressed the grooved surface for about 3 s, and then held for 10 s. The induced compression force was also kept as 20 N. Then, the loading bar and tactile sensor array slid along the grooved surfaces for about 40 s.

The sampling rate of the scanning circuit for the generated forces sensing was set as 0.1 kHz. For each experimental set, three repeated tests were carried out repeatability.

4. Results and Discussion

4.1. Slippage Detection in Sliding Movement

By using the developed FEM model, the sliding movement of the tactile sensor array over the flat surface was analyzed. During the simulation, the sliding movement is divided into two steps. Step I—compressing and holding: the tactile sensor array was vertically compressed against the flat surface until the total reaction force reaches up to about 15 N. This force was a little lower than that of experimental tests with the aim to improve the convergence property and the convergence effectiveness of sliding movement during FEM simulation. As for each sensing unit, the force difference between experimental and FEM simulation can be further decreased to about 0.56 N. Then this step was held for 1 s to maintain the normal force of 15 N. Step II—sliding: the sensor array was moved sliding along the x-axis direction at a speed of 1.0 mm/s. Because our developed tactile sensor array has extremely low weight less than 10 g, the moving speed lower than 1.0 mm/s can be considered as the quasi-static state movement. So, the moving speed lower than 1.0 mm/s has little effect on the FEM simulation predicted forces and will greatly reduce the simulation time.

FEM simulation results for each sensing unit at the end of the compression and sliding stages are shown in Figure 5. In Figure 5a,c, the generated normal stress along the z-axis is symmetrically distributed at the cross-section view of the sensing unit during the compression stage. During the sliding stage, the compression stress in the left region (marked as L) is generally increased and becomes greater than that in the right region (marked as R), as shown in Figure 5b,d. This phenomenon has been confirmed by other studies [11,12] and can be attributed to the torque caused by the friction between the flat plate and the PDMS bump. The shear deformation of the sensing unit also occurred during sliding movement. This can be explained as the localized displacement occurred at the contact region of the PDMS bump [14,15]. The lower boundary of the hemisphere-shaped bump cannot immediately follow the movement of the upper part of the tactile sensor due to the friction in the contact region. Thus, gross slippage will be generated as the distance between the contact region and the PET substrate is stretched long enough to overcome the effects of friction.

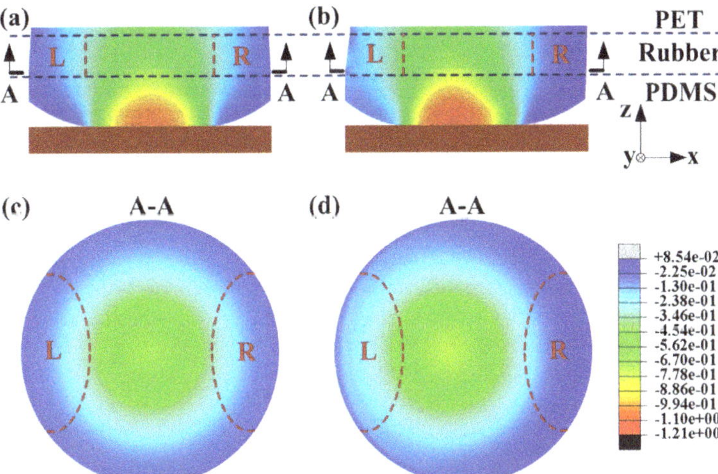

Figure 5. The stress distribution in the cross-section view of the sensing unit at the end of (**a**) compress and (**b**) sliding. A-A cross-section at the end of (**c**) compress and (**d**) sliding.

The normal force generated at the left and right areas of the sensing unit are extracted based on the simulation results, as shown in Figure 6a. In Figure 6a, the sliding direction of the tactile sensor array along the flat surface can be observed as from the red area to the blue area. The normal forces obtained at both the left and right area are gradually increased from zero to 0.07 N at the loading stage. During the sliding stage, the normal force at the left area is increased significantly, while the generated normal force in the right area is decreased. Though the force's amplitude was generally small, the relative change rate almost reaches 100%. Therefore, the sliding direction can be identified based on the obtained normal force curves in the left and right areas.

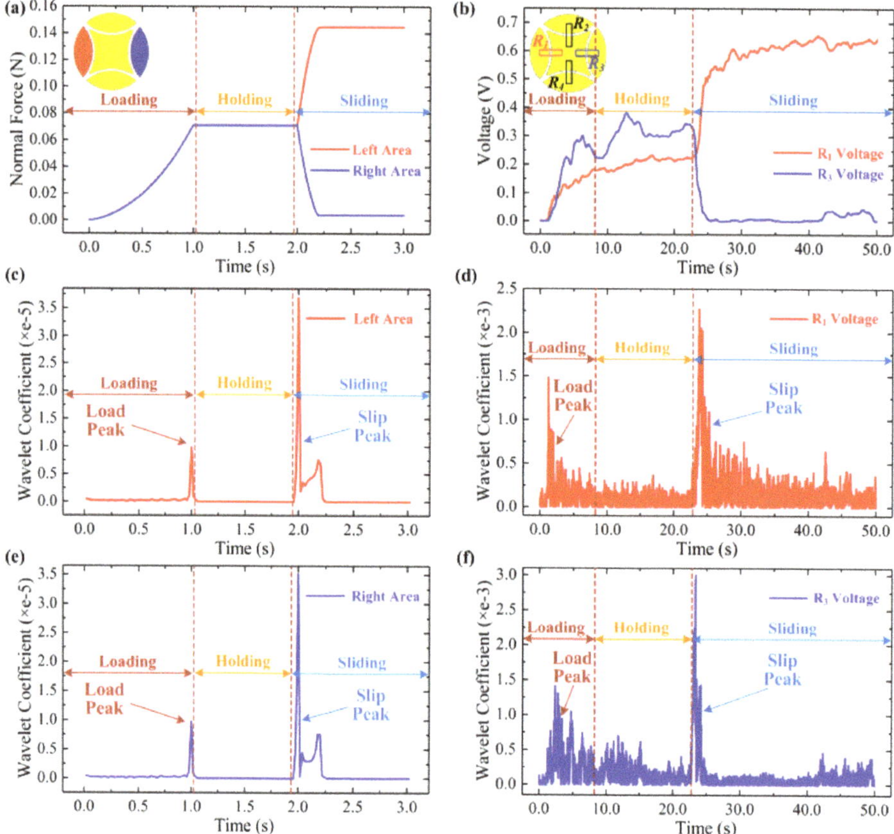

Figure 6. (**a**) Simulated normal force extracted from the left and right area of the patterned electrodes in the sensing unit (**b**) Measured voltages of R_1 and R_3 when sliding along a flat surface, DSWT analysis of the simulated normal force in left area (**c**) and right area (**e**) DSWT analysis of the measured voltages in R_1 (**d**) and R_3 (**f**).

To validate the FEM prediction, experimental tests are conducted. Figure 6b shows the measured voltages of the tactile sensor array in R_1 and R_3 resistors when the sensor array is compressing and sliding on a flat surface. The experimental setup and procedure adopted are presented in the preceding Section 3. Generally, the measured voltages of R_1 and R_3 have greater variations while having almost the same trends as that of the simulated normal force, as shown in Figure 6a. This is because the sensitivity of the utilized conductive rubber material in tactile sensor is over 500 kΩ /N when the applied force is lower than 0.7 N [12]. At the sliding stage, the measured voltage of R_1 is also increased, and the voltage of R_3 is decreased. Therefore, we can clearly distinguish the sliding occurrence and direction from either the FEM simulated normal forces or the measured voltages at the side electrode area and resistors.

Initial slippage detection is proven to be important for robotic hand grasping. Wavelet transform has been utilized to analyze the measured forces or voltages of the tactile sensors and demonstrates the ability to identify the change of the derivation for initial slippage discrimination [28]. Here, we also utilized the wavelet transform to analyze the simulated normal forces and measured voltages in Figure 6a,b. Figure 6c–f shows the discrete sequence wavelet transform (DSWT) results of the simulated normal forces and measured voltages of R_1 and R_3, respectively. We picked Coiflet as the mother wavelet function and set its length equal to 6. Two peaks can be observed at the transition moments from the loading to holding and from the holding to sliding stages, as shown in Figure 6c–f. The variations of the wavelet coefficient at initial sliding are much greater than that after loading. Thus, by setting a reasonable threshold value for the wavelet coefficient, the initial slippage can be distinguished as the tactile sensor array contacts and slides along object surfaces. More details of this method for slippage detection can be found in our previous study [11].

4.2. Surface Texture Recognition

4.2.1. Phase Delay Algorithm for Surface Texture Recognition

Spectrum analysis of the tactile sensor's output voltages can be used to determine the spatial periodical information of the utilized grooved surfaces. The spatial period value (D), defined as the distance between two adjacent ridges (as shown in Figure 7b), can be calculated as:

$$D = v/\text{MAF} \tag{3}$$

where v is the sliding speed, and MAF stands for the frequency with the maximum amplitude.

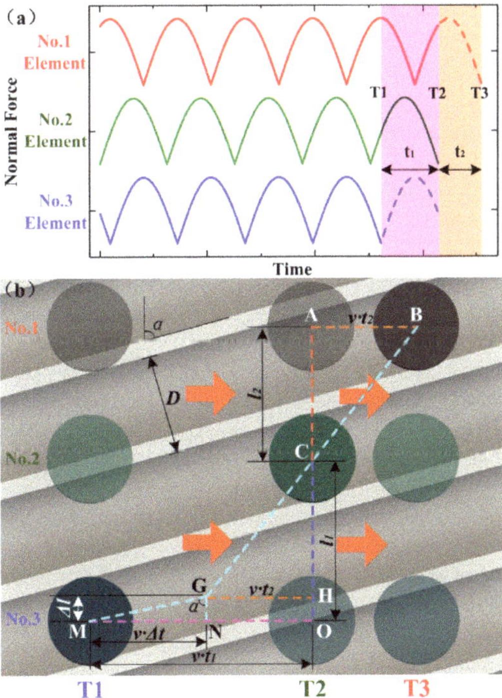

Figure 7. (a) Force-time curves of No. 1–3 units, (b) schematic view for the calculation of inclined angle α.

In practical application, the grooved texture in the plate surface usually has an inclined angle of α. Applying FFT to the obtained data may get us a fake spatial period value equal to D/cos(α). Therefore, the inclined angle (α) also needs to be determined. For this purpose, we changed the distance between the sensing units in each row and column to create different phase delays, as shown in Figure 7b. We assumed that the force-time curves of one column of three sensing units are as shown in Figure 7a. The horizontal movement ends at T_2, when No. 2 unit is at the center of the groove, which causes the force-time curve to end at a minimal value. At T_1, the No. 3 unit will also be in a similar position where it sustains the lowest pressure. If the sliding movement continues, the force of No. 1 unit will drop to the same value at T_3. In this condition, the movement path distances (AB and MO) can be calculated as vt_1 and vt_2, respectively, where t_1 and t_2 are the gaps between each moment, as indicated in Figure 7a. If the distance AC (l_1) or CO (l_2) is longer than that of D/sin(α), the inclined angle (α) can be calculated using the anti-trigonometric function and can be expressed as

$$\alpha = \arctan(\frac{v \cdot t_1}{l_1}) \tag{4}$$

The schematic diagram to calculate α is shown in Figure 7b. △CGH is first created as the same as △CBA, where point H is located on line CO. Then, we connect point G and point M and create a right triangle △GNM, where line GN is perpendicular to line MN. The angle ∠MGN is the same as that of inclined angle α. The length of GN and MN can be calculated as (l_1−l_2) and v·(t_1−t_2), respectively. Thus, the inclined angle α can be calculated as

$$\alpha = \arctan[\frac{v \cdot (t_1 - t_2)}{l_1 - l_2}] \tag{5}$$

Also, the spatial period value (D) can be calculated as

$$D = v \cdot \cos(\alpha)/\mathrm{MAF} \tag{6}$$

The procedure and flow chart of the phase delay algorithm for the grooved surface texture recognition is shown in Figure 8. The whole procedure can be mainly divided into three modules. In Slippage Judging module, the threshold based on wavelet coefficient is used. As the coefficient value is larger than 0.002, the program will jump out of the first loop and enter the Data Preprocessing module. The scanning circuit will sample the voltage data from three electrodes in the same column for 10 s. Using the low-pass filtering, the time gap (t_1 and t_2) is calculated based on the cross-correlation function analysis in this step. As for simulation, the force curves usually have an approximate sinusoidal shape. As for real tests, the measured voltage signals will be affected by external noises and vibrations, making the voltage curves not as good as that of the simulation results. Thus, we reconstructed the sine function of the characteristic frequency in the time domain and input these new curves into the cross-correlation function as shown in Figure 8. This step can be regarded as "band-pass filtering", which can improve the accuracy of the final results. After getting the reconstructed sine function, the program enters the Period Calculating module. By using Equations (5) and (6), the inclined angle (α) and spatial period value (D) can be calculated. Both maximum and minimum points of the cross-correlation curve will be taken as two different inputs. Therefore, we can obtain two spatial period results in the end of this module. If the difference between these two values is greater than 0.01 mm, the program will jump back and sample another set of data for a new round of calculation. Otherwise, the mean value will be output.

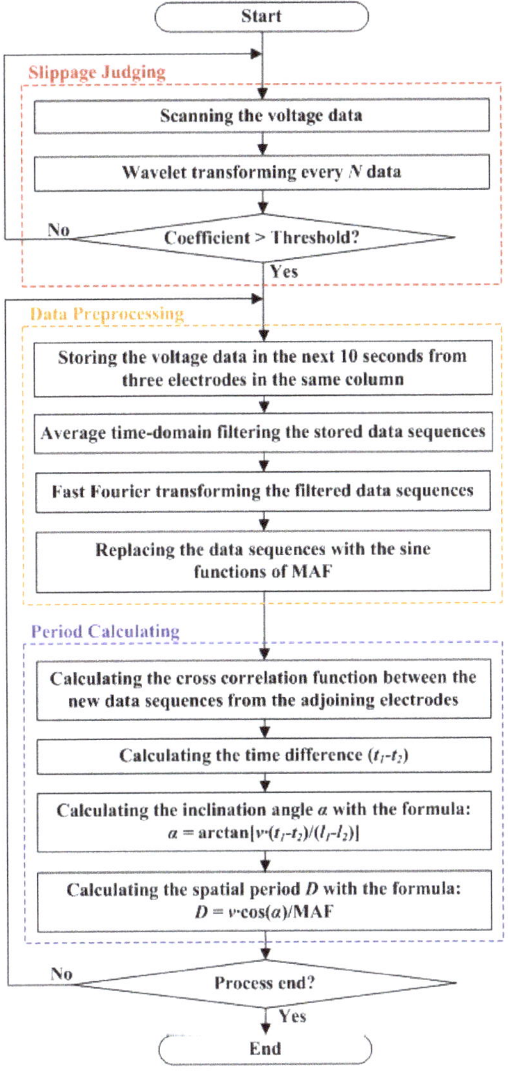

Figure 8. Flow chart and procedure to calculate the inclined angle and spatial period value for the grooved surfaces.

4.2.2. Spatial Period Discrimination

To verify the ability of the developed method for surface texture recognition, both simulation, and experimental tests were performed when the tactile sensor array slid along the grooved surfaces. For grooved surface recognition, the compression force was set as 15 N and 20 N for FEM simulation and experimental tests, respectively. The applied force during experimental tests is a little larger. The reasons are the measured output voltage of our tactile sensor array usually contains some noises from the scanning circuit and environment. If the applied force is too small, it will affect the accuracy of tactile sensor for surface texture recognition. Even the misjudging of the frequency characteristics may be occurred, and leads to false result. For the utilized grooved surfaces, as shown in Figure 9a, the inclined angle of the grooves on the plate is set as zero and spatial period as 0.9, 1.2 and 1.5 mm, respectively. As the sensor array slid along the surface, the measured voltages of one sensing unit for these grooved surfaces are shown in Figure 9b. We can see that the variation of voltage signals increases with the increase in the spatial period. For example, the grooved surface with 1.5 mm spatial period has the highest variation in the voltage curve. It is because narrower groove makes the compressed bumps of sensor less released.

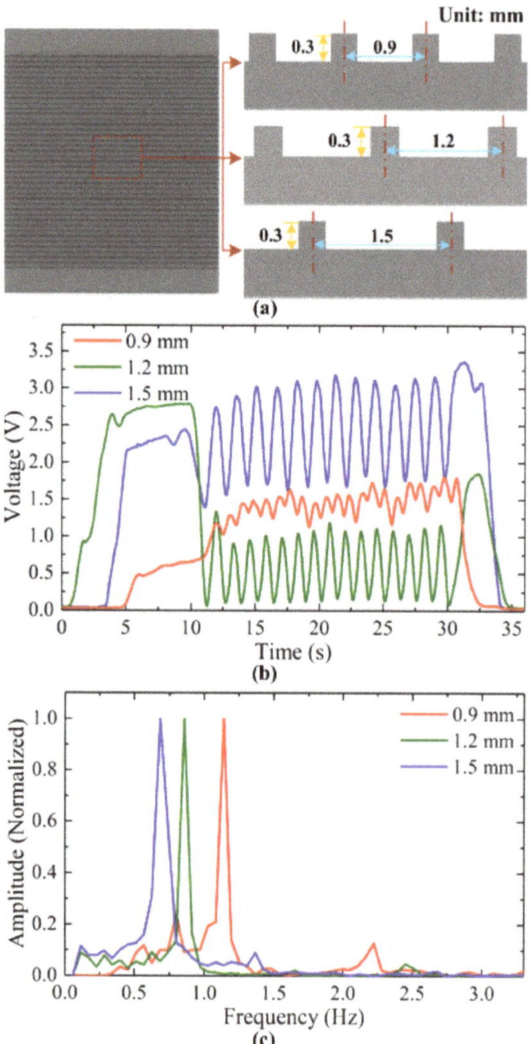

Figure 9. (**a**) Three grooved surfaces with different spatial periods of 0.9, 1.2 and 1.5 mm; (**b**) Measured voltage; (**c**) Spectrum analysis for surface texture recognition.

Using the calculation procedure in Figure 8 (simplified, as there is no inclined angle), spectrum analysis is conducted, as shown in Figure 9c. For comparison, the MAF and spatial periods of the grooved surfaces in simulation and experimental tests are calculated and listed in Table 2. We can see that the deviation of the calculated MAF and the spatial period from the real values are generally low as the errors are usually smaller than 6.7%. Thus, we can conclude that plate surfaces with different grooved textures can be recognized successfully using the method developed and the proposed tactile sensor array.

Table 2. Results of calculated spatial period and inclined angle in grooved surfaces.

	Spatial Period Discrimination (angle = 0°)				Inclined Angle Calculation (D = 1.2 mm)					
	Real D/mm	**0.90**	**1.20**	**1.50**	**Real Angle**	**0°**	**15°**	**30°**	**45°**	**60°**
Simulation	MAF | Error (Hz)	1.11 | 0.0%	0.78 | 6.0%	0.67 | 0.0%	α | Error	0.00° | 0.0%	11.51° | 23.3%	29.12° | 2.9%	44.10° | 2.0%	62.53° | −4.2%
	D | Error (mm)	0.90 | 0.0%	1.12 | 6.7%	1.50 | 0.0%	D | Error (mm)	1.14 | 5.0%	1.31 | −9.2%	1.16 | 3.3%	1.14 | 5.0%	1.23 | −2.5%
Experiment	MAF | Error (Hz)	1.06 | 4.5%	0.83 | 0.0%	0.67 | 0.0%	α | Error	0.48° | /	13.59° | 7.4%	29.90° | 0.3%	45.47° | 1.0%	61.50° | −2.5%
	D | Error (mm)	0.95 | −5.6%	1.20 | 0.0%	1.50 | 0.0%	D | Error (mm)	1.25 | −4.2%	1.22 | −1.7%	1.30 | −8.3%	1.31 | −9.2%	1.19 | 0.8%

4.2.3. Inclined Angle Calculation in Grooved Surfaces

To validate the developed method for the calculation of grooved surface with an inclined angle, we set l_1 and l_2 as 3.5 mm and 3.8 mm, respectively. For simulation and experimental tests, the grooved surface's spatial period is set as 1.2 mm and the inclined angle of the grooved patterns as 0°, 15°, 30°, 45° and 60°, respectively.

According to the calculation procedure and Equations (5) and (6), the inclined angle and spatial period for simulation and validation tests are calculated as shown in Table 2. We can see that both the calculated inclined angle and spatial period values for the simulation and experiments generally match well with the pre-determined results. For angle calculation, the greatest errors (23.3% for the simulation and 7.4% for the experiment) occur when the inclined angle equals 15°. It is because the slope of the inversed tangent function is extremely steep when the inclined angle is lower than 20°. Thus, even a small difference in the inputs would affect the accuracy of the calculated results. For the angle in the range of 30° to 60°, the relative errors are greatly reduced and less than 5.0%, as shown in Table 2. For the spatial period calculation, as it is also influenced by the sampling time, the error's variation shows a different trend compared with the previous one. Still, the biggest error is less than 9.2% for both simulation and experiment tests. Therefore, a plate with inclined arranged grooves on the surface can also be recognized using the developed method.

Typically, the simulated force curves for the inclined angle of 0°, 30° and 60° are shown in Figure 10a,c,e. The peak number in the force curve is decreased from 7 to 6 when the inclined angle increased from 0° to 30°. The peak number dramatically drops to 3 when the angle is raised up to 60°. This is due to the inclined grooves will enlarge the horizontal gap distance between the adjacent grooves. Under a uniform sliding motion, the larger angle will increase the time period, and in turn leads to less peaks, as show in Figure 10a,c,e. These three figures also show the trends that the overlapped force curves are gradually apart from each other. The green one is much closer to the red one than that of blue one. This phenomenon verifies that the existence of phase delay which caused by the inclined arranged grooves. Longer distance between elements 2 and 3 makes the phase delay more obviously. As in Figure 10e, the variation of the force curves drops from 0.08 N to 0.04 N with the increase of inclined angle. The curves stand for the total normal force applied on a fusiform area as shown in Figure 5. The inclined grooves arrangement will increase the minimum contacted area between the ridges and the fusiform area during sliding. Thus, the bump could not be fully released, and making the force curve variation become smaller.

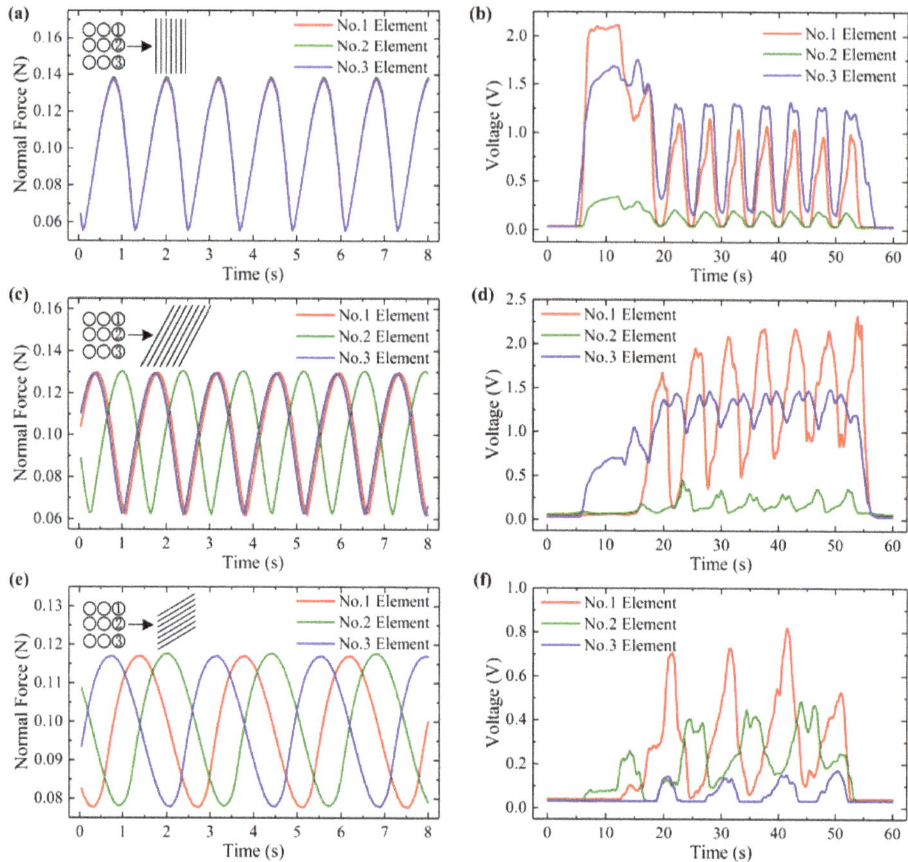

Figure 10. Simulated normal force curves when the sensor array is sliding along the grooved surface with an inclined angle α equal to (**a**) 0°, (**c**) 30° and (**e**) 60°. Measured voltages from the validation test for inclined angle calculation when α equals (**b**) 0°, (**d**) 30°, (**f**) 60°.

The measured voltage curves in real tests are shown in Figure 10b,d,f. Most phenomena discussed in the simulation cases could be verified here, like the peak number drops from 8 to 4 when the angle increases to 60°. The above results indicated that the grooved surfaces with different spatial period and inclined angle arrangement could be successfully discriminated by using the proposed flexible tactile sensor sliding motion and phase delay algorithm, thus may have potential in robotic grasping tasks for surface texture recognition.

5. Conclusions

This study develops a methodology using FEM modeling and the Phase Delay Algorithm to validate the flexible tactile sensor array for slippage and surface texture recognition in sliding motions. The structure and working principle of the tactile sensor array and its 3D FEM modeling are presented. The hyper-elastic Yeoh model is utilized to describe the material properties of PDMS, RTV adhesive, and conductive rubber utilized in the tactile sensors. For the sensor array sliding along the flat surface, both FEM simulation and experiments demonstrated that slippage occurrence and sliding direction can be determined based on the simulated normal force and measured voltages in the side resistors' region. For surface texture recognition, the Phase Delay Algorithm and its calculation procedure are developed. Results also showed that the grooved surface with and without inclined arranged grooves can be successfully discriminated.

This study opens up the opportunity to study surface texture identification for a flexible tactile sensor array in real applications. Optimal structural design of the flexible tactile sensor array including electrode pattern's design needs to be performed in future work. Further, the approach of using phase delay algorithm and artificial neural network for the developed tactile sensor array for robotic hand discrimination of unknown surface textures will also be conducted.

Author Contributions: Sensor design and modeling, Y.W.; FEM simulation and experimental tests, J.C.; conceptualization, D.M.

Funding: This work was supported by the National Natural Science Foundation of China (51575485), Zhejiang Provincial Funds for Distinguished Young Scientists of China (LR19E050001), and Creative Research Groups of National Natural Science Foundation of China (51821093).

Conflicts of Interest: The authors declare no conflict of interest.

References

1. Dahiya, R.S.; Metta, G.; Valle, M.; Sandini, G. Tactile sensing—from humans to humanoids. *IEEE Trans. Robot.* **2010**, *26*, 1–20. [CrossRef]
2. Chortos, A.; Liu, J.; Bao, Z. Pursuing prosthetic electronic skin. *Nat. Mater.* **2016**, *15*, 937–950. [CrossRef]
3. Damian, D.; Martinez, H.; Dermitzakis, K. Artificial ridged skin for slippage speed detection in prosthetic hand application. In Proceedings of the 2010 IEEE/RSJ International Conference on Intelligent Robots and Systems, Taipei, Taiwan, 18–22 October 2010; pp. 904–909.
4. Youref, H.; Boukallel, M.; Althoefer, K. Tactile sensing for dexterous in-hand manipulation in robotics—A review. *Sens. Actuator A Phys* **2011**, *167*, 171–187.
5. Feix, T.; Bullock, I.; Dollar, A. Analysis of Human Grasping Behavior: Object Characteristics and Grasp Type. *IEEE Trans. Haptics* **2014**, *7*, 311–323. [CrossRef] [PubMed]
6. Pritchard, E.; Mahfouz, M.; Iii, B.; Eliza, S.; Haider, M. Flexible capacitive sensors for high resolution pressure measurement. In Proceedings of the 2008 IEEE Sensors, Lecce, Italy, 26–29 October 2008; pp. 1484–1487.
7. Lee, H.; Chung, J.; Chang, S.; Yoon, E. Real-time measurement of the three-axis contact force distribution using a flexible capacitive polymer tactile sensor. *J. Micromech. Microeng.* **2011**, *21*, 35010–35018. [CrossRef]
8. Khan, S.; Tinku, S.; Lorenzelli, L.; Dahiya, R. Flexible tactile sensors using screen-printed p(vdf-trfe) and mwcnt/pdms composites. *IEEE Sens. J.* **2015**, *15*, 3146–3155. [CrossRef]
9. Cerveri, P.; Quinzi, M.; Bovio, D.; Frigo, C. A Novel Wearable Apparatus to Measure Fingertip Forces in Manipulation Tasks Based on MEMS Barometric Sensors. *IEEE Trans. Haptics* **2017**, *10*, 317–324. [CrossRef] [PubMed]
10. Yuan, Z.; Zhou, T.; Yin, Y.; Cao, R.; Li, C.; Wang, Z. Transparent and flexible triboelectric sensing array for touch security applications. *ACS Nano* **2017**, *11*, 8364–8369. [CrossRef] [PubMed]
11. Wang, Y.C.; Xi, K.L.; Mei, D.Q.; Liang, G.H.; Chen, Z.C. A flexible tactile sensor array based on pressure conductive rubber for contact force measurement and slip detection. *J. Robot. Mech.* **2016**, *28*, 378–385. [CrossRef]
12. Wang, Y.C.; Wu, X.; Mei, D.Q.; Zhu, L.F.; Chen, J.N. Flexible tactile sensor array for distributed tactile sensing and slip detection in robotic hand grasping. *Sens. Actuator A Phys.* **2019**, *297*, 111512. [CrossRef]
13. Zhang, Y.; Duan, X.; Zhong, G.; Deng, H. Initial slip detection and its application in biomimetic robotic hands. *IEEE Sens. J.* **2016**, *16*, 7073–7080. [CrossRef]
14. Ho, V.; Hirai, S. Two-dimensional dynamic modeling of a sliding motion of a soft fingertip focusing on stick-to-slip transition. In Proceedings of the 2010 IEEE International Conference on Robotics and Automation, Anchorage, AK, USA, 3–7 May 2010; pp. 4315–4321.
15. Ho, V.; Hirai, S. Three-dimensional modeling and simulation of the sliding motion of a soft fingertip with friction, focusing on stick-slip transition. In Proceedings of the 2011 IEEE International Conference on Robotics and Automation, Shanghai, China, 9–13 May 2011; pp. 5233–5239.

16. Dao, D.; Toriyama, T.; Wells, J.; Sugiyama, S. Micro force-moment sensor with six-degree of freedom. In Proceedings of the 2001 International Symposium on Micromechatronics and Human Science, Nagoya, Japan, 9–12 September 2001; pp. 93–98.
17. Youssefian, S.; Rahbar, N.; Torres-Jara, E. Contact behavior of soft spherical tactile sensors. *IEEE Sens. J.* **2014**, *14*, 1435–1442. [CrossRef]
18. Shimojo, M. Mechanical filtering effect of elastic cover for tactile sensor. *IEEE Trans. Robot. Autom.* **1997**, *13*, 128–132. [CrossRef]
19. Ho, V.; Nagatani, T.; Noda, A.; Hirai, S. What can be inferred from a tactile arrayed sensor in autonomous in-hand manipulation? In Proceedings of the 2012 IEEE International Conference on Automation Science and Engineering (CASE), Seoul, Korea, 20–24 August 2012; pp. 461–468.
20. Ayyildiz, M.; Güçlü, B.; Yildiz, M.; Basdogan, C. An Optoelectromechanical Tactile Sensor for Detection of Breast Lumps. *IEEE Trans. Haptics* **2013**, *6*, 145–155. [CrossRef] [PubMed]
21. Oddo, C.; Beccai, L.; Felder, M.; Giovacchini, F.; Carrozza, M. Artificial roughness encoding with a bio-inspired MEMS-based tactile sensor array. *Sensors* **2009**, *9*, 3161–3183. [CrossRef] [PubMed]
22. Oddo, C.; Controzzi, M.; Beccai, L.; Cipriani, C.; Carrozza, M. Roughness encoding for discrimination of surfaces in artificial active-touch. *IEEE Trans. Robot.* **2011**, *27*, 522–533. [CrossRef]
23. Fishel, J.; Loeb, G. Bayesian exploration for intelligent identification of textures. *Front. Neurorobot.* **2012**, *6*, 1–20. [CrossRef] [PubMed]
24. Liang, G.; Mei, D.; Wang, Y.; Dai, Y.; Chen, Z. A micro-wires based tactile sensor for prosthesis. In Proceedings of the the 6th International Conference on Intelligent Robotics and Applications, Busan, Korea, 25–28 September 2013; pp. 475–483.
25. Standard Test Methods for Rubber Properties in Compression. *ASTM D575–91(2012)*; ASTM International: West Conshohocken, PA, USA, 2012.
26. Yeoh, O.H. Some Forms of the Strain Energy Function for Rubber. *Rubber Chem. Technol.* **1993**, *66*, 754–771. [CrossRef]
27. Dogru, S.; Aksoy, B.; Bayraktar, H.; Alaca, B. Poisson's ratio of PDMS thin films. *Polym. Test.* **2018**, *69*, 375–384. [CrossRef]
28. Teshigawara, S.; Tadakuma, K.; Ming, A.; Ishikawa, M.; Shimojo, M. High sensitivity initial slip sensor for dexterous grasp. In Proceedings of the 2010 IEEE International Conference on Robotics and Automation, Anchorage, AK, USA, 3–7 May 2010; pp. 4867–4872.

Article

How the Skin Thickness and Thermal Contact Resistance Influence Thermal Tactile Perception

Congyan Chen * **and Shichen Ding**

School of Automation, Southeast University, Nanjing 210096, China; 220161453@seu.edu.cn
* Correspondence: chency@seu.edu.cn; Tel.: +86-138-1588-0379

Received: 25 December 2018; Accepted: 24 January 2019; Published: 25 January 2019

Abstract: A few experimental studies on thermal tactile perception have shown the influence of the thermal contact resistance which relates to contact surface roughness and pressure. In this paper, the theoretical influence of the skin thickness and the thermal contact resistance is studied on the thermal model describing the temperature evolution in skin and materials when they come in contact. The thermal theoretical profile for reproducing a thermal cue for given contact thermal resistance is also presented. Compared to existing models of thermal simulation, the method proposed here has the advantage that the parameters of skin structure and thermal contact resistance in target temperature profiles can be adjusted in thermal perception simulation according to different skin features or surface roughness if necessary. The experimental results of surface roughness recognition were also presented.

Keywords: thermal tactile perception; surface roughness; skin thickness; thermal perception reproduction

---

1. Introduction

Thermal perception is a rich, emotive, and entirely silent information source. For example, when our hands touch objects, thermal perceptions can provide information about their thermal characteristics, and help us recognize materials [1]. More, it could be used as an alternative mobile feedback channel when required, as it is silent for quiet environments, especially in electromagnetic interference case where monitors or headsets cannot work normally.

To simulate exchanging information by thermal tactile, some thermal displays have been developed for the reproduction of the thermal perception when a finger is in contact with a virtual or a remote real object [2,3]. Different thermal properties make different thermal profiles which result in different thermal tactile perceptions [4].

The relationship between the contact temperature evolution and the thermal characteristics have been studied to develop thermal feedback systems [5,6]. The works presented theoretical and experimental study of a model of heat exchange during hand-object interactions, and particularly evaluated by comparing the theoretical values of temperature changes to those experimentally measured [7,8]. The authors also studied how the contact area and contact pressure during hand-object interactions affected the skin temperature changes.

When a finger contacts a material, not only the thermal characteristics of the skin and the material but also the skin physical structure and their contact state have an influence on the heat exchange during hand-object interactions. There are some significant factors affecting the heat exchange.

The work [9] proposed a model for heat transfer occurring between the finger skin and any given surface based on an electrical analogy, and discussed the comparison between the model and some experimental results by considering various phenomena like the applied pressure by the finger, the speed of the blood circulation, the interface state. The experimental results [10] have shown that there was a small change in skin temperature as a function of the surface roughness of the contact material. Some analyses of the relation between the skin temperature change and contact pressure in a thermal display have been also presented with the aid of an infrared thermal measurement system [11]. A quantified model for local and overall thermal sensations in non-uniform thermal environments is also proposed in [12].

However, some features still should be specified. For example, what influence do the finger skin physical characteristics (The thickness of skin, the surface roughness) exert on thermal perception during hand-object interactions?

The thermoreceptors which function as thermal sensors are scattered between the dermis and epidermis [13]. The thermal perception originates from the temperature drop and its change rate at thermoreceptors, which relates to physical and thermal properties, initial temperature difference, thermal contact resistance, and other factors [14].

Finite-element calculations have been applied to simulation in the case of thermal contact resistance in order to simulate flat and smooth surfaces of objects with different properties in virtual reality [2]. The fingertip surface roughness was measured and the thermal contact resistance of the finger was estimated based on an infrared camera during interaction phases to study the influence of surface properties on thermal tactile perception [15]. With the addition of thermal contact resistance to the thermal model, the temperature profiles of the skin and materials become more realistic.

However, the role of a fingertip skin thickness and the roughness of contact surface, which influences the temperature drop at thermoreceptors, should be investigated more. The authors in [13] have considered the thickness of the epidermis and dermis to model the temperature responses at the cold receptors. The thermal model was studied from the consideration that the skin is not regarded as homogeneous but as consisting of three layers of tissue that differ in terms of heat flux: epidermis, dermis, and endodermis [16].

The skin thickness and the roughness of contact surface were both considered in the present paper in order to study the thermal responses of the fingertip as it makes contact with materials with different surface roughness. We believe it is helpful in modelling thermal tactile perception with considering these factors.

2. Thermal Modelling

The condition of contact between a fingertip and a material is shown in Figure 1. The fingertip has three layers: the inner layer is well known as the subdermal zone; the middle layer is the dermis; and the outer layer is the epidermis. The thermal contact resistance is specifically considered here.

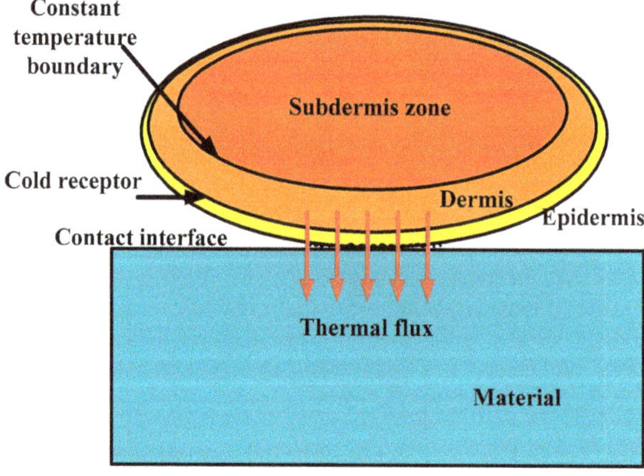

Figure 1. The condition of contact between a fingertip and a material.

As the subdermal zone is with large heat capacity and low heat conductivity, its temperature will remain stable normally when the outside thermal stimulus changes. A thermoreceptor is a sensory receptor that helps us get "cold" or "hot" sensations. The temperature and its change rate at warm or cold thermoreceptors are the main source of cold or hot sensation [17].

Cold thermoreceptors are much more numerous than warm thermoreceptors by a ratio of up to 30:1, and respond to the decreases in temperature over a temperature range of 5–43 °C [18,19]. The different depths of cold thermoreceptor may bring about different temperature drops. However, the actual depth is dependent on the certain thickness of skin.

The thermal exchange between the skin and a material in contact with is a transient process and is dominated by heat conduction. Normally heat flows from the skin to materials. And the thermal contact resistance between the fingertip and materials is involved unavoidably in real contact. So, it should be also considered in thermal modelling.

These total thermal resistances result in less temperature drop at thermoreceptors than that in theory. For above considerations, an experimental system has been constructed to measure the surface roughness [8]. And a device was also developed to implement the skin-object thermal contact resistance measurement [20].

The whole thermal contact system with thermal resistance considered can be modelled as shown in Figure 2.

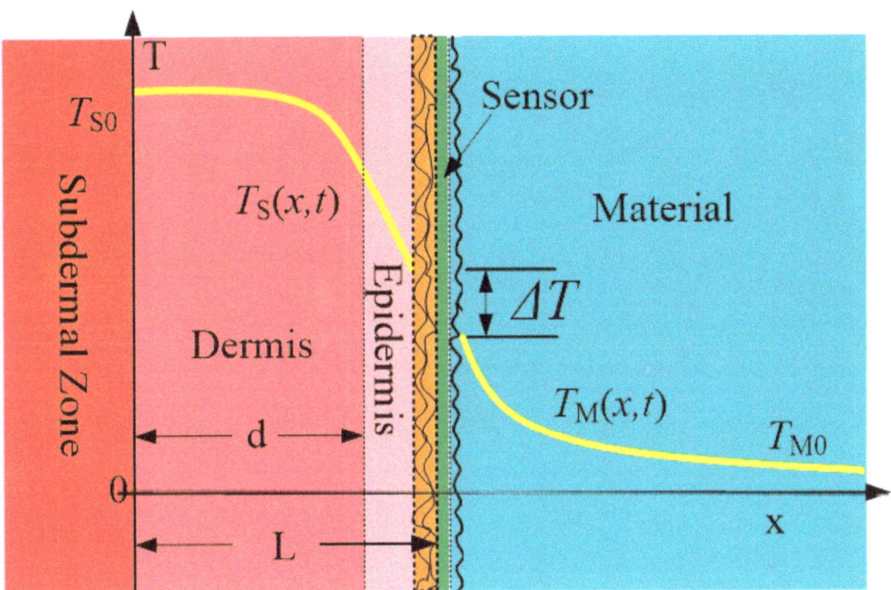

Figure 2. Thermal contact system.

$T_S(x,t)$ and $T_M(x,t)$ refer to finger temperature and material temperature, and T_{S0} and T_{M0} are their initial values. Here L is the thickness of skin, and d is the thickness of its dermis. And there is a thermal sensor between the finger and material to measure the interface temperature. Here, the thermal resistance which results in some temperature drop ΔT is denoted as R here.

Let λ_i be the thermal conductivity, c_i be the specific heat, and ρ_i be the density, then $\alpha_i = \frac{\lambda_i}{\rho_i c_i}$ is known as the thermal diffusion coefficient, and $\beta_i = (\lambda_i \rho_i c_i)^{\frac{1}{2}}$ is the thermal contact coefficient, here

$i = S$ and M represents skin and material respectively. So, the governing Equations of the skin and material can be expressed as:

$$\frac{\partial\theta_S(x,t)}{\partial t} = \alpha_S\frac{\partial^2\theta_S(x,t)}{\partial x^2}, \left\{\begin{array}{l}\theta_S(x,t) = 1, t = 0\\ \theta_S(x,t) = 1, x = 0\end{array}\right\} \tag{1a}$$

$$\frac{\partial\theta_M(x,t)}{\partial t} = \alpha_M\frac{\partial^2\theta_M(x,t)}{\partial x^2}, \left\{\begin{array}{l}\theta_M(x,t) = 0, t = 0\\ \theta_M(x,t) = 0, x \to \infty\end{array}\right\} \tag{1b}$$

where the relative residual temperature is defined:

$$\theta_i(x,t) = \frac{T_i(x,t) - T_{M0}}{T_{S0} - T_{M0}}, \ i = S, M.$$

When $t > 0$, the boundary conditions are $\theta_S(L^-,t) = \theta_M(L^+,t) + \Delta\theta$ at the contact interface, and $-\lambda_S\frac{\partial\theta_S(x,t)}{\partial x}|x = L^- = -\lambda_M\frac{\partial\theta_M(x,t)}{\partial x}|x = L^+ = \frac{\theta_S(L^-,t)-\theta_M(L^+,t)}{R}$, $t > 0$. By introducing Laplace transforms to Equations (1a) and (1b), they are transformed to the differential Equations in variable x:

$$\frac{d^2\bar{\theta}_S(x,s)}{dx^2} = \frac{1}{\alpha_S}\left[s\bar{\theta}_S(x,s) - 1\right] \tag{2a}$$

$$\frac{d^2\bar{\theta}_M(x,s)}{dx^2} = \frac{1}{\alpha_M}s\bar{\theta}_M(x,s) \tag{2b}$$

where s is the common Laplace complex variable.

Let $\mu = \sqrt{\frac{\alpha_S}{\alpha_M}}$, $\beta = \frac{\lambda_S}{\lambda_M\mu} = \frac{\beta_S}{\beta_M}$, $H_1 = \frac{\beta+1}{R\lambda_S}$, $H_2 = \frac{\beta-1}{R\lambda_S}$, $\gamma = \frac{\beta-1}{\beta+1} = \frac{H_2}{H_1}$ and take the inverse Laplace transform of Equations (2a) and (2b). The approximate solution of the temperature in skin and material can be gotten as (3a) and (3b) according to the Laplace transform table in [21].

$$\theta_S(x,t) = 1 - \frac{\beta_M}{\beta_S+\beta_M}\sum_{n=0}^{\infty}(-1)^n\gamma^n\left\{erfc\left[\frac{(2n+1)L-x}{2\sqrt{\alpha_S t}}\right] - erfc\left[\frac{(2n+1)L+x}{2\sqrt{\alpha_S t}}\right] - e_{SR}\right\} \tag{3a}$$

$$\theta_M(x,t) = \frac{\beta_S}{\beta_S+\beta_M}\sum_{n=0}^{\infty}(-1)^n\gamma^n\left\{erfc\left[\frac{2nL-\mu(x-L)}{2\sqrt{\alpha_S t}}\right] + erfc\left[\frac{(2n+2)L+\mu(x-L)}{2\sqrt{\alpha_S t}}\right] - e_{MR}\right\} \tag{3b}$$

where $e_{SR} = e^{H_1(L-x)+H_1{}^2\alpha_S t}erfc\left[H_1\sqrt{\alpha_S t} + \frac{L-x}{2\sqrt{\alpha_S t}}\right] - e^{H_1(L+x)+H_1{}^2\alpha_S t}erfc\left[H_1\sqrt{\alpha_S t} + \frac{L+x}{2\sqrt{\alpha_S t}}\right]$, $e_{MR} = e^{H_1\mu(x-L)+H_1{}^2\alpha_S t}erfc\left[H_1\sqrt{\alpha_S t} + \frac{\mu(x-L)}{2\sqrt{\alpha_S t}}\right] - e^{H_1[2L+\mu(x-L)]+H_1{}^2\alpha_S t}erfc\left[H_1\sqrt{\alpha_S t} + \frac{2L+\mu(x-L)}{2\sqrt{\alpha_S t}}\right]$.

The temperatures in the fingertip skin and the material can be written as respectively:

$$T_S(x,t) = T_{S0} - (T_{S0} - T_{M0})\frac{1-\gamma}{2}\sum_{n=0}^{\infty}(-1)^n\gamma^n\left\{erfc\left[\frac{(2n+1)L-x}{2\sqrt{\alpha_S t}}\right] - erfc\left[\frac{(2n+1)L+x}{2\sqrt{\alpha_S t}}\right] - e_{SR}\right\} \tag{4a}$$

$$T_M(x,t) = T_{M0} + (T_{S0} - T_{M0})\frac{1+\gamma}{2}\sum_{n=0}^{\infty}(-1)^n\gamma^n\left\{erfc\left[\frac{2nL-\mu(x-L)}{2\sqrt{\alpha_S t}}\right] + erfc\left[\frac{(2n+1)L+\mu(x-L)}{2\sqrt{\alpha_S t}}\right] - e_{MR}\right\}. \tag{4b}$$

It is evident that the temperature drop at the thermoreceptors is dependent on not only T_{S0}, T_{M0}, β_S, β_M, α_S, α_M, but also L, d, and the thermal resistance R. Here β_S and α_S are the thermal parameters characterizing the skin of fingertip, L and d are the physical parameters characterizing the skin of fingertip, and R is the thermal resistance parameter characterizing the contact state for thermal conductivity.

Before simulating using the above thermal model, the thermal contact resistance can be given from the following relation [15]:

$$R_c = \frac{0.8R_q}{\lambda_{SF}\Delta a}\left(\frac{H}{P}\right)^{0.95} \tag{5}$$

where $\lambda_{SF} = \frac{2\lambda_S \cdot \lambda_a}{\lambda_S + \lambda_a}$ (W/m·k) is the harmonic mean thermal conductivity of the contact interface, $R_q = \left[R_S^2 + R_a^2\right]^{0.5}$ is the effective root mean square surface roughness, Δa is the effective absolute average surface asperity slope, $H = 12.5$ g/mm^2 is the skin microhardness, and P is the contact pressure. Here some parameters were adopted directly in the following simulation from the reference [15]: $\Delta a = 0.3$, $H = 12.5$ g/mm^2, $R_S = 21.69$ μm, and R_a is for the surface roughness of material. The contact pressure is given with a contact force of 2 N and contact area of 135 mm^2.

The theoretical temperature evolutions of stainless steel (SS) for different thicknesses of epidermis are shown in Figure 3. As denoted above, L is the thickness of fingertip skin, and d is the thickness of dermis, then $L - d$ is the thickness of epidermis. Here the initial temperatures of finger and material are set as 36 °C and 20 °C, respectively. The thermal characteristics applied in the simulation have been listed in Table 1. The simulation results show that the temperature evolutions at thermoreceptors are quite different for different thicknesses of epidermis.

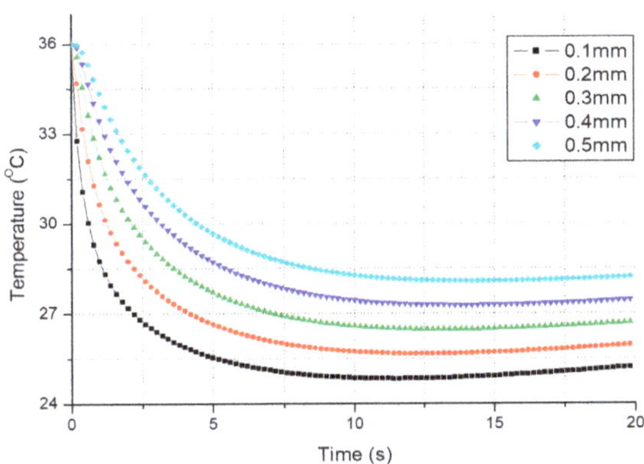

Figure 3. The temperature evolutions at thermoreceptors for different thicknesses of epidermis (Material = SS with absolutely smooth surface). SS—stainless steel.

Table 1. Thermal properties of stainless steel (SS) and skin.

Material	λ (W/(m·K))	ρ (kg/m³)	c (J/(kg·K))	β (W·s$^{1/2}$/(m²·K))
SS	14.9	7900	447	7253.71
Skin	0.34	1200	3340	1167.36

The terms e_{SR} and e_{MR} are factors for thermal contact resistance. Comparing with the model without considering thermal contact resistance, the factor e_{SR} will bring a temperature drop loss at thermoreceptors:

$$|\Delta T_S(x,t)| = \frac{\beta_M}{\beta_S + \beta_M} e_{SR} |T_{S0} - T_{M0}|. \tag{6}$$

For the same material, the temperature drop at thermoreceptors depends on not only the initial temperature difference between skin and the material but also their thermal contact state. The temperature drop will reduce with the increase of the thermal contact resistance.

The theoretical temperature evolutions at thermoreceptors with different surface roughness of stainless steel (SS) are illustrated in Figure 4a, where each temperature drop and its change rate reduce obviously with the increase of the surface roughness. And in Figure 4b, the steady-state temperature (replaced with the value at 20th second in the simulation) at thermoreceptors becomes higher with the increase of surface roughness for the heat flux conducted out of the skin becomes less.

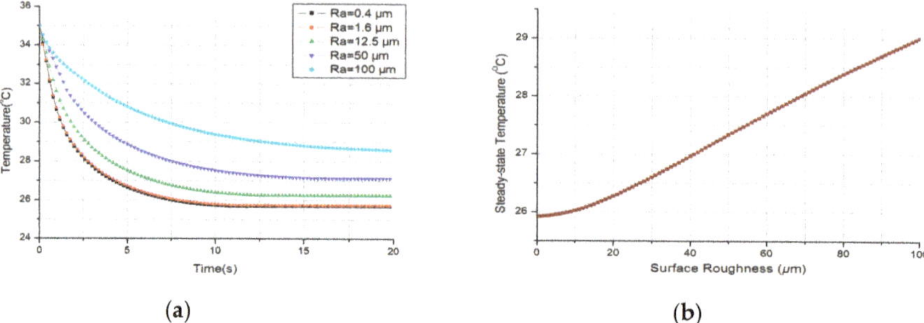

(a) (b)

Figure 4. The simulations for different surface roughness of SS: (**a**) The temperature evolutions at thermoreceptors; (**b**) The relationship between surface roughness and steady-state temperature at thermoreceptors.

It is shown in Figure 4b that the different surface roughness results in different steady-state temperatures at thermoreceptors. Namely the absolute temperature drops for the same material are also different after first several seconds of contact. And this may bring some different degrees of thermal tactile perception.

Remark 1. *From Figure 3, besides the initial temperature difference* $|T_{S0} - T_{M0}|$ *between skin and material and their thermal properties, one factor influencing temperature evolution at thermoreceptors is the thickness of skin. The thicker the skin, the smaller is the temperature drop at thermoreceptors.*

Skin thickness varies considerably according to the race, age, sex and region of the body surface. For example, for Korean population, the thickness of epidermis varies from 31 μm to 637 μm, while the thickness of dermis varies from 469 μm to 1942 μm [22]. So, the factor of skin thickness should be studied in thermal simulation.

Remark 2. *Another factor is the thermal contact resistance. It is obvious in Figure 4a that each temperature evolution at thermoreceptors is quite different from others as the contact surface roughness of each material sample is not the same. So, does each steady-state temperature drop. Both of the theory and simulation results show that the temperature drop at the thermoreceptors becomes less with the increase of the thermal contact resistance. So, the influence of thermal contact resistance should be also considered in modelling for thermal tactile perception.*

Now consider how to simulate the thermal perception of a given material with the initial temperature T_{M0} when considering the influence of both skin thickness and thermal contact resistance.

To reproduce an appointed thermal cue via a thermal tactile display, the contact temperature should be controlled to track the corresponding target temperature profile. A thermal sensor was situated between the fingertip skin and the material to measure the interface temperature. The theoretical profile of the interface temperature can be gotten from the following Equation:

$$T_M(x,t)|L^+ \leftarrow x = T_{M0} + (T_{S0} - T_{M0})\frac{1+\gamma}{2}\left\{erfc\left[\frac{\mu(x-L)}{2\sqrt{\alpha_S t}}\right] + erfc\left[\frac{2L+\mu(x-L)}{2\sqrt{\alpha_S t}}\right] - e_{MR}\right\}|L^+ \leftarrow x. \quad (7)$$

Equation (7) derives from Equation (4b) when x approaches L^+. $T_M(x,t)|L^+ \leftarrow x$ denotes the theoretical interface temperature of material side. It should be noted that the thermal contact resistance can be adjusted in thermal perception simulation. The theoretical interface temperature profiles for different surface roughness of SS are illustrated in Figure 5.

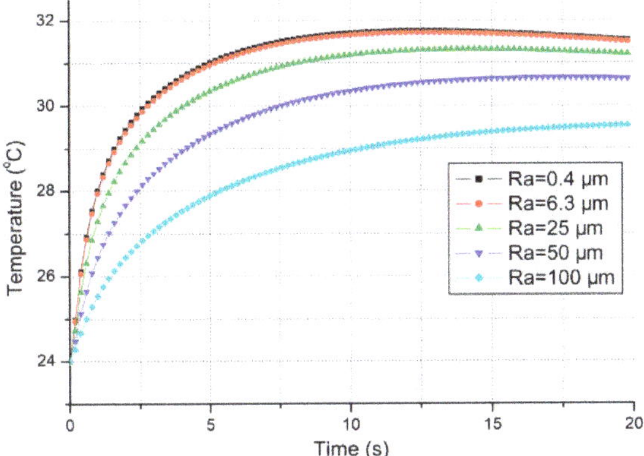

Figure 5. The theoretical interface temperature profiles for different surface roughness (material = SS, $R_s = 21.69$ μm).

With the increase of surface roughness, the thermal contact resistance increases and the heat flux out of the skin becomes less, and the steady-state temperature drop becomes less. So, the surface roughness of material also results in some difference in thermal tactile perception.

From the definition in [4], the theoretical relative recognizing profiles with thermal resistance at thermoreceptors can be gotten as:

$$\psi(t) = \frac{T_{S0} - T_S(d,t)}{T_{S0} - T_{M0}} = \frac{1+\gamma}{2}\left\{erfc\left[\frac{L-d}{2\sqrt{\alpha_S t}}\right] - erfc\left[\frac{L+d}{2\sqrt{\alpha_S t}}\right] - e_{SR}\right\} \quad (8)$$

In ideal case, the relative recognizing profiles with the same thermal characteristics are consistent for different initial temperatures. However, in real case, the thermal contact resistance has some influence on relative recognizing profile as discussed above. The relative recognizing profiles of SS with different surface roughness are illustrated in Figure 6.

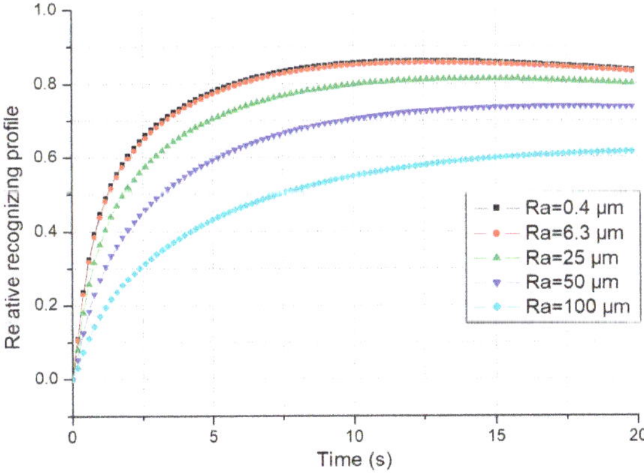

Figure 6. The relative thermal recognizing profiles (material = SS, $R_s = 21.69$ μm).

Remark 3. *It is feasible to apply relative thermal recognizing profiles to thermal tactile perception simulation with adjusting the parameter of thermal contact resistance. Due to inter-individual variations, it is difficult to set the initial skin temperatures exactly or thermal contact resistances to specific ones. However, we can manage* *to measure them and provide some approximate values. As soon as they are determined, the theoretical profiles are then gotten, and the corresponding thermal cues can be reproduced by a thermal tactile display.*

3. Experiments

The two experiments have been presented here. The first experiment is designed to study the influence of the thermal contact resistance on the thermal recognizing profile. The second one is aimed to verify the influence of the skin thickness and investigate the difference between real and simulated surface roughness recognition.

3.1. Influence of the Surface Roughness on the Thermal Recognizing Profiles

In order to evaluate the influence of thermal contact resistance on thermal modelling, an experiment was designed to investigate the evolution of interface temperature for different surface roughness of the same material. One set of temperature profile was measured for different surface roughness of SS.

The object with different surface roughness was standard samples processed by surface shot-peening, whose surface sizes are 20 mm × 23 mm or 50 mm × 46 mm with the thickness of 3 mm. As is shown in Figure 7, it is divided into 8 parts of different surfaces roughness. The object is made of nickel alloy using precision electroforming technology and has an advantage of high hardness, good abrasion resistance and good rust prevention. It makes this experiment safe and reliable.

Figure 7. The samples with different surface roughness (material = SS).

The participant cleaned his right hand before starting the experiment. A platinum thin-film thermal sensor (polyimide with foil backing, Minco S651) was glued to the fingertip of the right index finger with a biocompatible cyanoacrylate to decrease thermal resistance of contact.

The initial temperature of skin was about 35 °C, warmed and maintained with a hot-water bag beforehand. The room temperature was maintained at 24 °C, and also measured by a thermal sensor whenever necessary. The participant was instructed to place his right index fingertip on each sample in turn. The contact time for each trial lasted more than 25 s. However, the change of the contact temperature was only recorded in first 25 s for the transient process is over. Every temperature evolution was measured by the sensor, and illustrated in Figure 8.

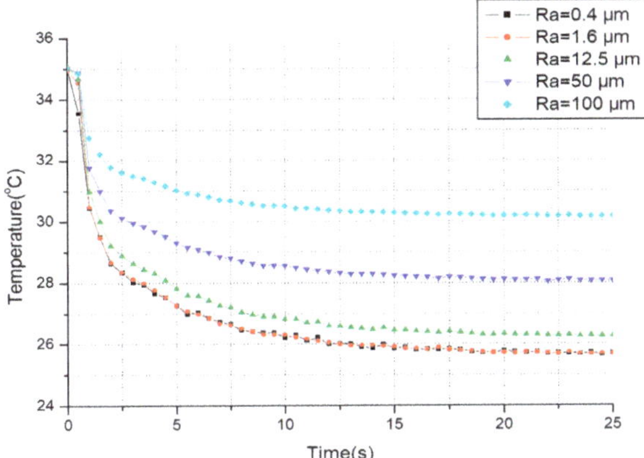

Figure 8. The experimental temperature evolution for different surface roughness (material = SS).

The experimental temperature evolutions in Figure 8 are similar to those in Figure 4. With the increase of the surface roughness of SS, the steady-state temperature drop (approximately replaced by the values at 25th second) also becomes less. This verifies that the surface roughness of SS affects the thermal tactile perception during hand-object interactions. The perception will become weaker because of the increase of the thermal resistance.

The comparisons between simulation and experimental data at 20 s are shown in Table 2. There are slight differences between simulation and experiment. Due to the limited sensor size and the spherical surface of the roughness samples (50 μm and 100 μm), the actual contact pattern including contact area and the contact pressure is different from others by degrees, resulting in a large deviation from the theoretical value.

Table 2. Comparisons between simulation and experimental data.

Roughness (μm)	Simulation Data (°C)	Experiment Data (°C)	Errors (%)
0.4	25.704	25.776	0.280
1.6	25.729	25.751	0.086
12.5	26.269	26.340	0.270
50	27.338	28.116	2.84
100	28.820	30.021	4.17

3.2. Experiment of Recognition of Different Surface Roughness

The second experiment was designed to measure subject's ability to identify SS samples with different surface roughness, and investigate the difference between real and simulated surface roughness recognition. The participants are thirty normal healthy adults including twelve women and eighteen men aged between 18 and 45 years in experiments. They were all right-handed, but with different occupations, for example, student, teacher, worker, farmer, etc. Before the experiment, they were simply trained to contact samples with about 2 N pressure expertly. Besides, their skin thickness was estimated by calculating the dimension of the skin's bio-speckles [23].

According to their skin thickness, the participants are split two groups: group A and group B. Each group has fifteen participants. The group A has five men and ten women, whose skin thickness is smaller than 800 μm, 600–750 μm. The group B has twelve men and three women, whose skin thickness is greater than 800 μm, about 900–1200 μm.

The experimental material is SS, whose surface roughness is listed in Table 3.

Table 3. Roughness of experimental sample.

Roughness Number	Roughness Value (R_a, µm)
1#	1.6
2#	12.5
3#	50
4#	100

Each participant's index finger was first cleaned in order to not interfere with the contact. The participants' initial fingertip skin temperatures ranged from 35.5 °C to 36 °C, warmed beforehand with a hot-water bag, and measured by a radiation thermometer just before touching. The room temperature was maintained at about 20 °C. In experiment there was a Platinum thin-film thermal sensor (also S651) glued with thermally conductive silicone which can decrease the thermal contact resistance between the sensor and samples. The sensor also helps to prohibit the surface texture tactile perception.

At the beginning, four stainless steel samples with different surface roughness were shown and presented to thirty participants by thermal feedback to become familiar with them.

In the real surface roughness recognition experiment, four stainless steel samples with different roughness were presented to participants randomly with three repetitions of each sample. Participants were forbidden to watch the procedure. When making contact with a sample they were instructed to judge it. No feedback was given regarding the correctness of each judgment. The contact time for each trial was not more than 20 s. After finishing the real material recognition experiment, participants could touch the samples again and reviewed their thermal cues. The results were labeled into two group A and B with real material.

In the next simulated surface roughness recognition experiment, the theoretical thermal cues are reproduced from the Equation (7). Except the skin thickness, the physical properties of skin and contact state in thermal reproduction were chosen as the same in the above simulation for simplicity. The mean skin thickness was set as 650 µm for group A and 1000 µm for group B, respectively.

The thermal tactile display device applied to simulate the different thermal cues has been developed [24]. It consists of a Peltier pad, two thermal sensors for ambient temperature and contact temperature, a radiation thermometer for finger temperature and four pressure sensors for the force of touch. The maximum difference of temperature between the cold and hot sides is more than 80 °C.

The target temperature profiles were given by the theoretical temperature based on Equation (7). As long as the initial skin and material temperatures also their thermal characteristics were set respectively, then the corresponding target temperature profile was specified for the given surface roughness and skin thickness.

Four samples with different simulated roughness were presented to participants randomly with three trials. After hearing a sound cue, a participant put the right index finger on the sensor attached on the Peltier pad. At the same time, the temperature profile was set to the corresponding target one. The participants took their hands back to the hot-water bag after speaking out the choice of the simulated roughness. The contact time for each trial was also not more than 20 s.

The two groups' responses in term of the correct percentage of roughness recognition for both real and their corresponding simulated samples are illustrated in Figure 9.

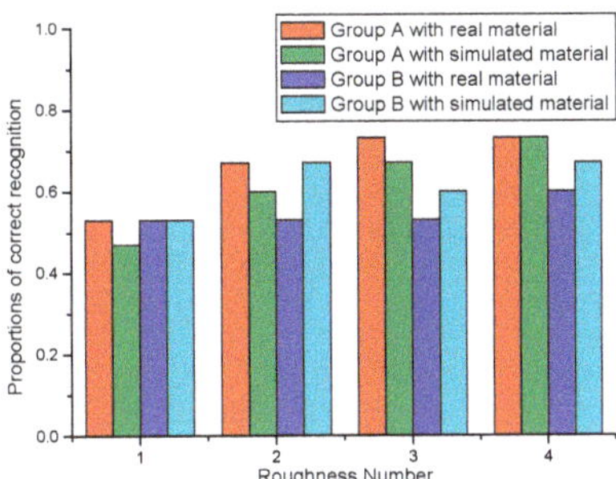

Figure 9. The proportions of correct recognition of two groups for real and simulated SS samples.

As shown in Figure 9, the correct proportion of surface roughness recognition goes up slowly with the difference of the thermal recognition profiles which become more obvious, as shown in Figure 8. However, the correct proportions of surface roughness recognition are still not quite high. The sample with surface roughness 1# (R_a = 1.6 μm) is often confused with other samples, both for the real and simulated samples. And the results also indicate that when only thermal cues are available, surface roughness recognition is quite not definite even when there is obvious difference of roughness.

On the other hand, comparing the correct recognition proportions of group A and group B for the same real surface roughness, the proportion of the group A is higher than that of the group B to a certain extent. This means that the skin thickness has some influence on thermal perception.

As pointed out above, thicker skin will bring about less temperature drop at thermoreceptors in contact with real material. For simulated material, however, this difference becomes less because the temperature drop loss is compensated according to our thermal model.

With the addition of thermal contact resistance to the thermal model, the temperature profiles of the skin and materials become more realistic. Although the different surface roughness can be recognized by the thermal tactile perception to a certain extent, the thermal perception is a complex process. It also relates to the participant's psychological response. However, the results show at least that the influence of surface roughness on thermal perception is existed basically in experiment.

Comparing to previous thermal simulation methods, the one proposed here has an advantage that the parameters of skin structure or surface roughness in target temperature profile can be adjusted in thermal simulation experiment according to different skin feature or material surface roughness if necessary.

4. Conclusions

This paper investigated the factors influencing thermal tactile perception based on thermoreceptors. Thickness of skin and thermal contact resistance have been considered in thermal modelling, and the theoretical temperature profiles in skin and material have been presented when they contact with each other. Also, a method has been proposed to reproduce thermal cues for different skin thickness and surface roughness. With the consideration of thermal contact resistance in thermal modelling, the temperature profiles of the skin and materials become more realistic. The experimental results of roughness recognition for real and simulated materials indicate that material roughness and the skin thickness influence the thermal perception.

Author Contributions: Conceptualization, C.C.; Data curation, C.C. and S.D.; Formal analysis, C.C.; Investigation, C.C.; Methodology, C.C. and S.D.; Resources, C.C.; Supervision, C.C.; Writing—original draft, C.C.; Writing—review & editing, S.D.

Funding: This research was funded by the National Natural Science Foundation of China (NSFC) (grant number U1531104 and U1731120) and the Special Fund of Jiangsu Province for Transformation of Scientific and Technological Achievements (grant number BA2017043).

Conflicts of Interest: The authors declare no conflict of interest.

References

1. Beccai, L.; Totaro, M. Editorial for the special issue on tactile sensing for soft robotics and wearables. *Micromachines* **2018**, *9*, 676. [CrossRef] [PubMed]
2. Citerin, J.; Pocheville, A.; Kheddar, A. A touch rendering device in a virtual environment with kinesthetic and thermal feedback. In Proceedings of the IEEE International Conference on Robotics and Automation, Orlando, FL, USA, 15–19 May 2006.
3. Drif, A.; Citerin, J.; Kheddar, A. Thermal bilateral coupling in teleoperators. In Proceedings of the IEEE/RSJ International Conference on Intelligent Robots and Systems, Edmonton, AB, Canada, 2–6 August 2005.
4. Chen, C.; Qiu, S. Influence of temperature profiles on thermal tactile sensation. *Adv. Robot.* **2014**, *28*, 53–61. [CrossRef]
5. Ho, H.; Jones, L. Material identification using real and simulated thermal cues. In Proceedings of the 26th Annual International Conference of the IEEE Engineering in Medicine and Biology Society, San Francisco, CA, USA, 1–5 September 2004.
6. Ho, H.; Jones, L. Development and evaluation of a thermal display for material identification and discrimination. *ACM Trans. Appl. Percept.* **2007**, *4*, 1–24. [CrossRef]
7. Guiatni, M.; Kheddar, A. Theoretical and experimental study of a heat transfer model for thermal feedback in virtual environments. In Proceedings of the IEEE/RSJ International Conference on Intelligent Robots and Systems, Nice, France, 22–26 September 2008.
8. Ho, H.; Jones, L. Modeling the thermal responses of the skin surface during hand-object interactions. *J. Biomech. Eng.* **2008**, *130*, 021005. [CrossRef] [PubMed]
9. Benali-Khoudja, M.; Hafez, M.; Alexandre, J.; Benachour, J.; Kheddar, A. Thermal feedback model for virtual reality. In Proceedings of the International Symposium on Micromechatronics and Human Science, Nagoya, Japan, 19–22 October 2003.
10. Galie, J.; Ho, H.; Jones, L. Influence of contact conditions on thermal responses of the hand. In Proceedings of the Third Joint Eurohaptics Conference and Symposium on Haptic Interfaces for Virtual Environment and Teleoperator Systems, Salt Lake City, UT, USA, 18–20 March 2009.
11. Ho, H.; Jones, L. Infrared thermal measurement system for evaluating model-based thermal displays. In Proceedings of the Second Joint EuroHaptics Conference and Symposium on Haptic Interfaces for Virtual Environment and Teleoperator Systems, Tsukaba, Japan, 22–24 March 2007.
12. Jin, Q.; Li, X.; Lin, D.; Shu, H.; Sun, Y.; Ding, Q. Predictive model of local and overall thermal sensations for non-uniform environments. *Build. Environ.* **2012**, *51*, 330–344. [CrossRef]
13. Ring, J.; Dear, R. Temperature transients: A model for heat diffusion through the skin, thermoreceptor response and thermal sensation. *Indoor Air* **1991**, *4*, 448–456. [CrossRef]
14. Park, M.; Bok, B.G.; Ahn, J.H.; Kim, M.S. Recent Advances in Tactile Sensing Technology. *Micromachines* **2018**, *9*, 321. [CrossRef] [PubMed]
15. Ho, H.; Jones, L. Thermal model for hand-object interactions. In Proceedings of the Symposium on Haptic Interfaces for Virtual Environment and Teleoperator Systems, Alexandria, VI, USA, 25–26 March 2005.
16. Deml, B.; Mihalyi, A.; Hanning, G. Development and experiment evaluation of a thermal display. In Proceedings of the 2006 EuroHaptics, Paris, France, 3–6 July 2006.
17. Boulais, N.; Misery, L. The epidermis: A sensory tissue. *Eur. J. Dermatol.* **2008**, *18*, 119–127. [PubMed]
18. Lv, Y.; Liu, J. Effect of transient temperature on thermoreceptor response and thermal sensation. *Build. Environ.* **2007**, *42*, 656–664. [CrossRef]
19. Jones, L.; Berris, M. The psychophysics of temperature perception and thermal-interface design. In Proceedings of the 10th Symposium on Haptic Interfaces for Virtual Environment & Teleoperator Systems, Orlando, FL, USA, 24–25 March 2002.
20. Saggin, B.; Tarabini, B.; Lanfranchi, G. A device for skin-contact thermal resistance measurement. *IEEE Trans. Instrum. Meas.* **2012**, *61*, 489–495. [CrossRef]
21. Ozisik, N. *Heat Conduction*, 2nd ed.; Wiley-Interscience: New York, NY, USA, 1993.
22. Lee, Y.; Hwang, K. Skin thickness of Korean adults. *Surg. Radiol. Anat.* **2002**, *24*, 183–189. [PubMed]

23. Wen, W.; Fu, R.; Ba, E.; Liu, Y.; Zhang, X.; Ma, S. An application of correction dimension of bio-speckles to measuring skin thickness. In Proceedings of the Technical Digest, CLEO/Pacific Rim '99, Pacific Rim Conference on Lasers and Electro-Optics (Cat. No.99TH8464), Seoul, Korea, 30 August–3 September 1999.
24. Chen, C.; Gu, X.; Li, Z.; Qiu, S. Thermal tactile display and its application for material identification. *ICIC Express Lett.* **2012**, *6*, 177–184.

MDPI
St. Alban-Anlage 66
4052 Basel
Switzerland
Tel. +41 61 683 77 34
Fax +41 61 302 89 18
www.mdpi.com

Micromachines Editorial Office
E-mail: micromachines@mdpi.com
www.mdpi.com/journal/micromachines

www.ingramcontent.com/pod-product-compliance
Lightning Source LLC
LaVergne TN
LVHW070643170726
843515LV00004B/316

* 9 7 8 3 0 3 9 3 6 5 0 1 2 *